62nd Air Traffic Control Association Annual Conference and Exposition 2017

Including Papers from 61st Air Traffic Control Association Annual Conference & Exposition 2016

National Harbor, Maryland, USA
15 – 18 October 2017

ISBN: 978-1-5108-5365-2

Printed from e-media with permission by:

Curran Associates, Inc.
57 Morehouse Lane
Red Hook, NY 12571

Some format issues inherent in the e-media version may also appear in this print version.

Printed by Curran Associates, Inc. (2018)

For permission requests, please contact Air Traffic Control Association
at the address below.

Air Traffic Control Association
1101 King St., Suite 300
Alexandria, Virginia 22314
USA

Phone: (703) 299-2430
Fax: (703) 299-2437

info@atca.org

Additional copies of this publication are available from:

Curran Associates, Inc.
57 Morehouse Lane
Red Hook, NY 12571 USA
Phone: 845-758-0400
Fax: 845-758-2633
Email: curran@proceedings.com
Web: www.proceedings.com

TABLE OF CONTENTS

Airborne Trajectory Management (ABTM):

A Blueprint for Greater Autonomy in Air Traffic Management

Capt. William B. Cotton
National Institute of Aerospace, Hampton, VA 23666

David Wing
NASA Langley Research Center, Hampton, VA 23681

Introduction

The aviation users of the National Airspace System (NAS) – the airlines, General Aviation (GA), the military and, most recently, operators of Unmanned Aircraft Systems (UAS) – are constrained in their operations by the design of the current paradigm for air traffic control (ATC). Some of these constraints include ATC preferred routes, departure fix restrictions and airspace ground delay programs. As a result, most flights cannot operate on their most efficient business trajectories and a great many flights are delayed even getting into the air, which imposes a significant challenge to maintaining efficient flight and network operations.

Rather than accepting ever more sophisticated scheduling solutions to accommodate the existing constraints in the airspace, a series of increasingly capable airborne technologies, integrated with planned improvements in the ground system through the Federal Aviation Administration (FAA) Next Generation Air Traffic Management System (NextGen) programs, could produce much greater operational flexibility for flight path optimization by the aviation system users. These capabilities, described in research coming out of NASA's Aeronautics Research Mission Directorate, can maintain or improve operational safety while taking advantage of air and ground NextGen technologies in novel ways. The underlying premise is that the nation's physical airspace is still abundant and underused, and that the delays and inefficient flight operations resulting from artificial structure in airspace use and procedural constraints on those operations may not be necessary for safe and efficient flight.

This article is not an indictment of today's NAS or the people who run it. Indeed, it is an exceptional achievement that Air Traffic Management (ATM) – the complex human/machine conglomeration of communications, navigation and surveillance equipment and the rules and procedures for controlling traffic in the airspace – has both the capacity and enables the degree of efficiency in air travel that it does. But it is also true that sixty years of the "radar religion" (i.e., reliance on radar-based command and control) has produced several generations of ATM system operators and researchers who believe that introducing automation within the existing functional structure of ATM is the only way to "modernize" the system. Even NextGen, which began as a proposal for "transformational" change in the way ATC is performed, has morphed over the last decade and a half to become just the inclusion of Global Positioning System (GPS) for navigation, Automatic Dependent Surveillance Broadcast (ADS-B) for surveillance, and Data Communications (Data Comm) for communications, while still operating in rigidly structured airspace with human controllers being responsible for separation and traffic flow management (TFM) within defined sectors of airspace, using the same horizontal separation standards that have been in use since raw primary radar was introduced in the 1950s.

No system as massive as the current NAS ATM can be replaced with a better system while simultaneously meeting the transportation and other aviation needs of the nation. A new generation of more flexible operations must emerge and yet coexist in harmony with the current operation (i.e., share the same airspace without segregation), thereby enabling a long-term transformation to take place in the way increasing numbers of flights are handled. Market forces will be the ultimate driver of this transformation, and investment realities mandate that real benefits must accrue to the first operators to adopt these new capabilities. In fact, the kinds of missions envisioned in the emerging world of UAS operations, unachievable under conventional ATM, demand that this transformation take place.

Airborne Trajectory Management (ABTM) is proposed as a series of transformational steps leading to vastly increased flexibility in flight operations and capacity in the airspace to accommodate many varied airspace uses while improving safety. As will be described, ABTM enables the gradual emergence of a new paradigm for user-based trajectory management in ATM that brings tangible benefits to equipped operators at every step while leveraging the air and ground investments of NextGen. There are five steps in this ABTM transformation.[1] NASA has extensively studied the first and last of these steps, and a roadmap of increasing capabilities and benefits is proposed for bridging between these operational concepts.

Step One - Traffic Aware Strategic Aircrew Requests (TASAR)

The roadmap begins with Traffic Aware Strategic Aircrew Requests (TASAR).[2] It is well known that today's ATC system permits greater flexibility in the choice of route and altitude for airborne aircraft than during the flight planning process. Flight plans must be filed to conform with established constraints in the airspace that are static, i.e., in place regardless of existing traffic. Once airborne, pilots routinely request shortcuts that are often accommodated by air traffic controllers who are doing their best to increase efficiency in the system, given the constraints of traffic flows and airspace structure to which they are required to adhere. Usually, this takes the form of a "direct to" clearance, eliminating a dogleg in the cleared route. But it is also true that "direct" is not the optimal path (i.e., shortest flight time or lowest fuel burn) unless the wind field is completely uniform, something that almost never occurs in nature. That is why modern flight planning systems seek the best wind route, optimizing the flight to minimize time, fuel burn, or a balanced combination of both. But flight planning systems must also modify the best wind route to conform to the constraints imposed by ATC, or else the filed flight plan won't be accepted into the FAA's EnRoute ATC Modernization (ERAM) computer. TASAR, depicted in Figure 1, goes a step beyond what can be accomplished pre-flight by the operator's flight planning systems by providing the pilot a real-time flight optimizing tool in the cockpit.

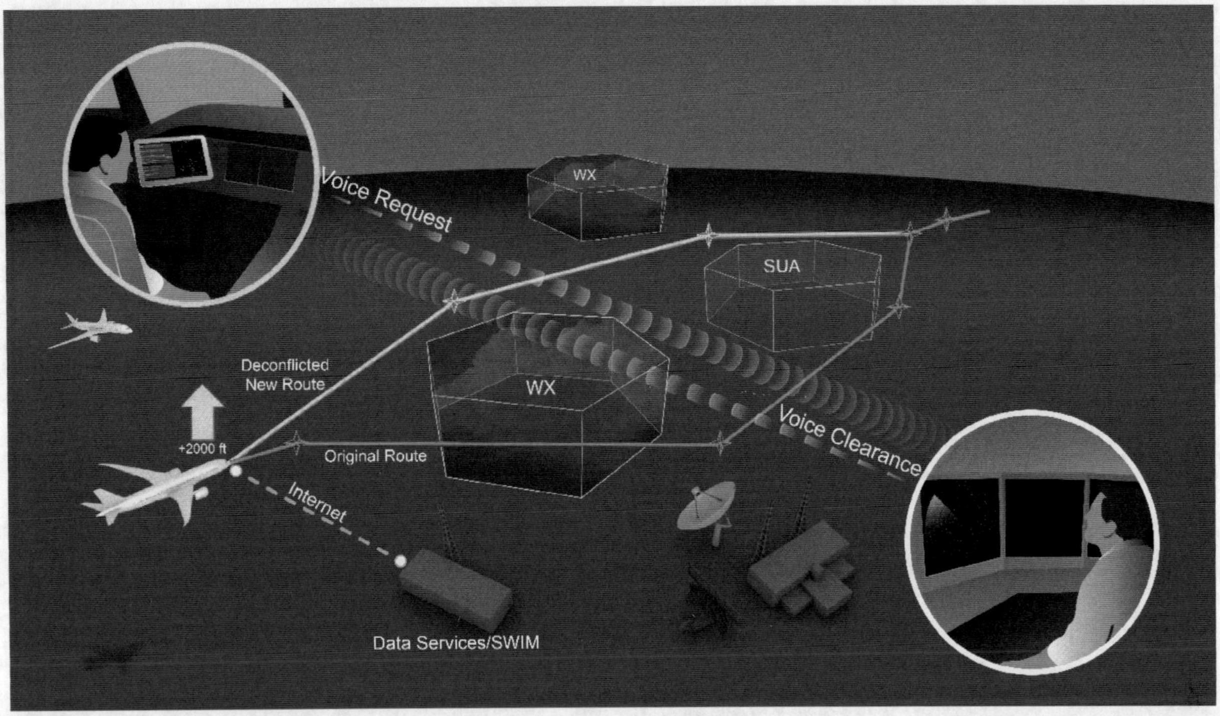

Figure 1 ABTM Roadmap Step 1 - TASAR

TASAR employs an EFB application with access to aircraft flight information from onboard avionics and airspace/environmental information on wind and weather forecasts, Special Use Airspace (SUA) activity, and other NAS constraints through the aircraft's broadband internet connection to a System Wide Information Management (SWIM) gateway or other third party service providers. This TASAR EFB application is exemplified by NASA's prototype Traffic Aware Planner (TAP) software application.[4] Using a sophisticated route optimization algorithm, TAP computes the best route and altitude from present position forward once each minute, using the most recent wind grid available, filtering out solutions that are conflicted by ADS-B IN traffic, active SUA, or hazardous weather, and presenting reroute solutions (lateral, vertical, and combination lateral/vertical) to the pilot. After review, the pilot may request the TAP-generated trajectory change from the controller, leveraging the greater rerouting flexibility while airborne to achieve significant operational benefits. Because the TAP solutions are already vetted for traffic, SUA, and weather, TASAR requests stand a better chance of being approved by ATC than a typically uninformed pilot request.

The development of the TASAR operational concept and the TAP software has taken place under NASA sponsorship for more than a decade and has been evaluated in simulations and flight tests by airline pilots and air traffic controllers. NASA has partnered with two airlines to conduct operational flight trials of TASAR beginning in the summer of 2017.

Step Two - Digital TASAR

Digital TASAR adds data communications to basic TASAR. TASAR's initial design accommodates the fact that requested changes in the flight trajectory can be cumbersome to describe in voice transmissions by constraining the number of requested new waypoints to no more than two, and these must be named waypoints. Similarly, no more than one altitude change is requested at a time. Procedurally, only one

TASAR request is made per control sector to keep the workload acceptable to both pilots and controllers. In Digital TASAR, depicted in Figure 2, Data Comm and the resultant graphic representation to ATC of the requested trajectory change mitigate these limitations of the voice request environment. Upgraded for Digital TASAR, the optimized solutions produced by the TAP software and their outcomes (i.e., fuel and/or time saved) will be loadable to the Flight Management System (FMS) and sent as a request to ERAM through the FAA's Data Comm Services with the push of a button, requiring neither voice requests nor manual keyboard entry. The request is then graphically displayed to the controller for consideration. Approval, amendment, or denial would also take place through the FAA's Data Comm Services.

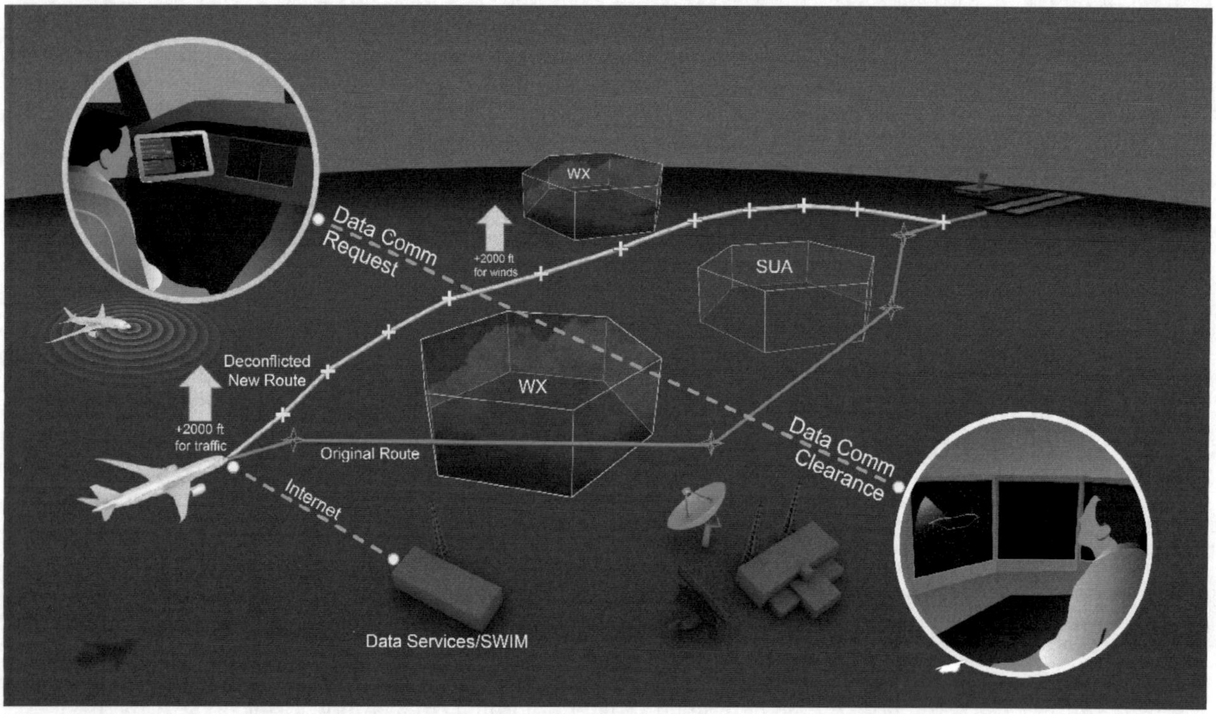

Figure 2 ABTM Roadmap Step 2 - Digital TASAR

Digital TASAR requests have higher value than basic TASAR because they are not constrained by the voice request and approval limitations, enabling more complex optimized trajectory change requests. However, they are actually simpler for the pilots and controllers to use because they are graphically portrayed and digitally shared without voice reference to individual waypoint names. Because the trajectory change *request* is not safety critical, unlike the amended clearance, a pre-Data Comm Full Services (circa 2022) method for down-linking the request can be considered. For example, oceanic aircraft have used the Future Air Navigation System (FANS) 1 system for decades for handling alpha numeric requests and re-clearances. On the aircraft, this will require a TAP connection to the data communications system with attendant certification considerations. Airborne internet access to ERAM through a gateway like the Leidos flight plan forwarding system provides another method for making Digital TASAR requests. Digital TASAR will also simplify coordination with the airlines' flight dispatch function, allowing dispatchers to provide real-time adjustments to the optimization objective used by TAP to reflect total airline network optimization considerations, such as passenger connections, aircraft scheduling, and maintenance routing.

Step Three - 4D TASAR

4D TASAR adds the fourth dimension of time and/or speed management to the TAP optimization algorithms and enables integrating the solutions with such systems as the airborne Flight Deck Interval Management (FIM) application and the ground-based Time Based Flow Management (TBFM) automation for arrivals to capacity constrained airports. This provides an additional opportunity to save time and fuel beyond the lateral and vertical trajectory optimizations made using a constant cruise Mach number. It enables dynamic control of the Cost Index by the airline and also permits the flight to be inserted in the correct sequence in the arrival flow without being placed in a physical line of Miles-in

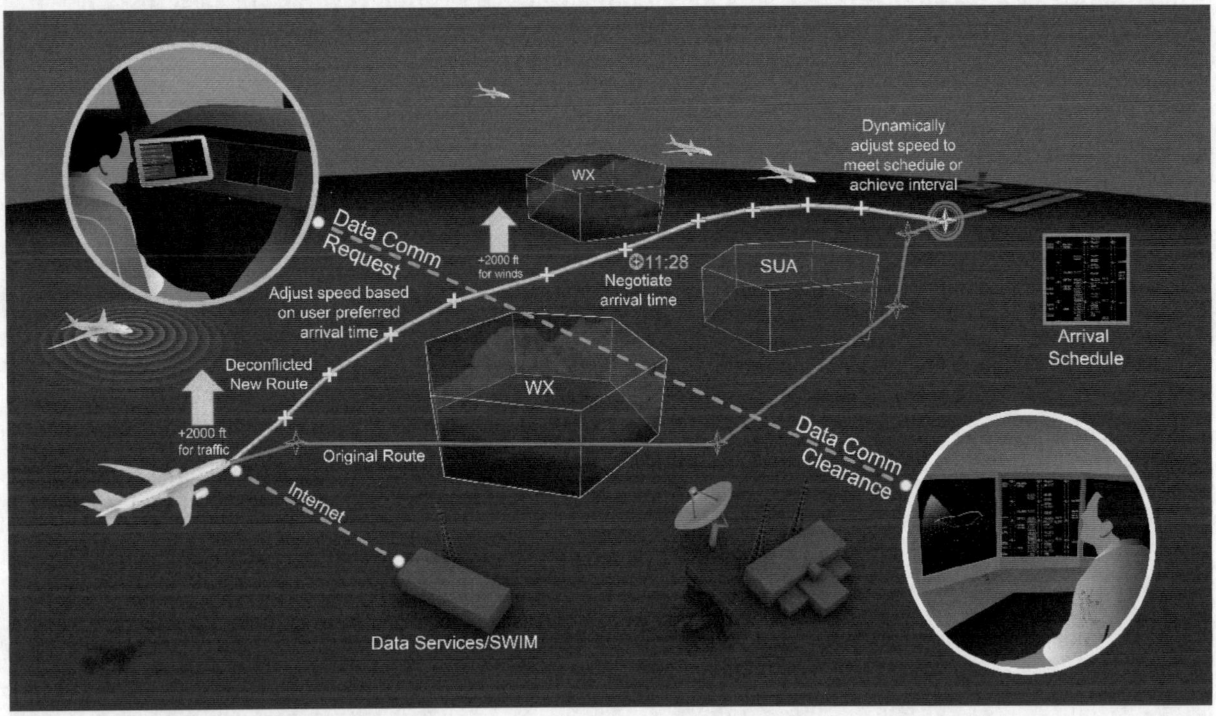

Figure 3 ABTM Roadmap Step 3- 4D TASAR

Trail traffic, a process that prevents every aircraft in the "conga line" from optimizing their speed.

The coordination of 4D TASAR (depicted in Figure 3) with the FAA's TBFM automation further reduces the chance for rejection of the trajectory optimization requests and strengthens the business case for using integrated avionics in the aircraft. In fact, the opportunity for full flight path optimization may provide a sufficient business case for upgrading aircraft for both FIM and Data Comm, including the higher level of certification required in this equipment. But the integration of the 4D tool with TBFM is only relevant at the largest airports where excess traffic demand presents a continuing operational constraint. For some airline flights, most GA flights, and nearly all UAS operations, major airport congestion is not an issue because they do not typically use these airports. For these users, the enhanced optimization capability provided by 4D TASAR can provide additional fuel and time savings, unimpeded by destination airport constraints.

Step Four - Strategic ABTM

In Strategic ABTM, depicted in Figure 4, trajectory changes will purposefully begin in the next sector (beyond the current sector) and will ensure no new conflicts are created in that sector (similar to a controller's responsibility when handing off aircraft). Such requests will be submitted to receive expedited (eventually automatic) approval from ATC without impacting the air traffic controller of the aircraft's current sector. When sent via Data Comm to the ERAM computer, the Strategic ATBM trajectory changes will automatically update the flight plan while simultaneously becoming the active route in the FMS on the aircraft. These changes (starting in the next sector) will not impact the controller in the active sector since they begin beyond his area of consideration and thus will not change the locus of responsibility for separation. ATC interventions for Strategic ABTM flights will be rare because the trajectory changes will be de-conflicted and preserve the essential constraints dictated by airspace use and TFM. While pilot assessment and approval is still required before sending Strategic ABTM solutions to ERAM, these TAP solutions will have become the primary mode of trajectory management in the cockpit. These iterated solutions will contain the latest in airline fleet and network optimization

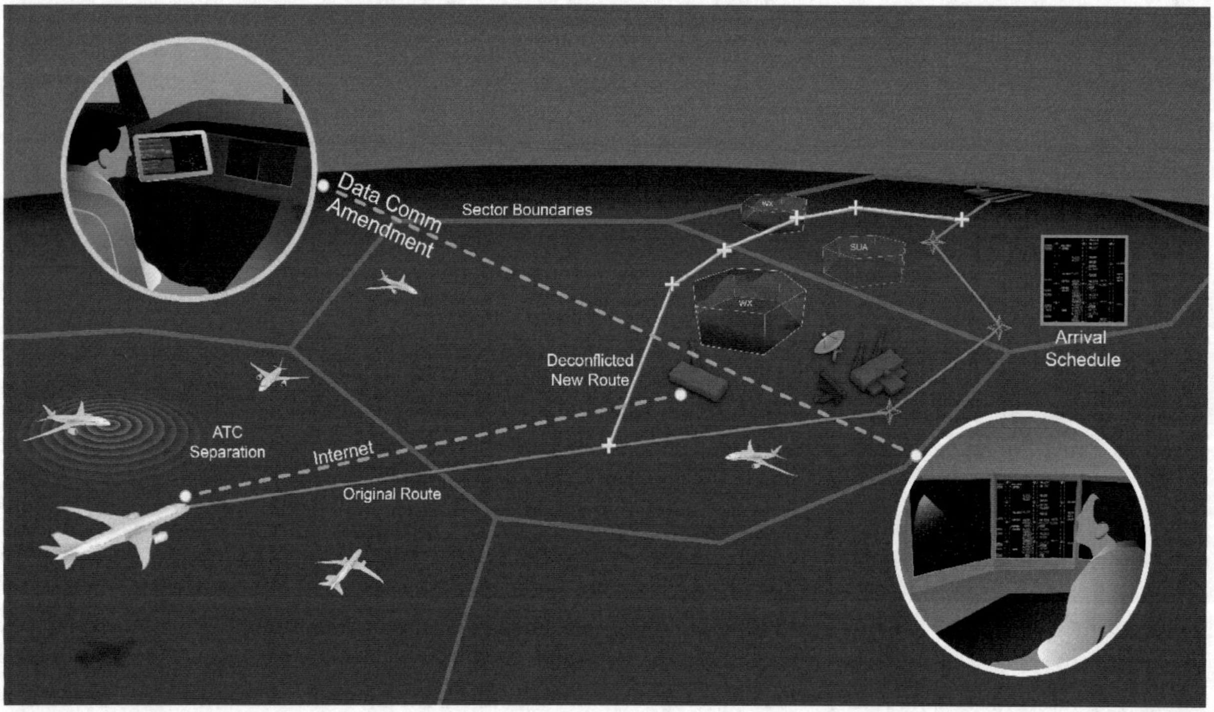

Figure 4 ABTM Roadmap Step 4 - Strategic ABTM

objectives, wind optimization, weather, traffic and airspace hazard avoidance.

During the years of experience gained in operating the previous versions of TASAR, the traffic de-confliction algorithm in TAP will have been refined and shown to provide reliably de-conflicted, optimized solutions. This extensive dataset will provide the foundation for the final step of ABTM wherein the separation function can be provided by the onboard trajectory management system. Traffic de-confliction is part of the legacy of TAP because it is software initially created for NASA's extensive research in airborne self-separation. The predecessor tool, NASA's Autonomous Operations Planner

(AOP),[5] has both a strategic (longer range) and a tactical component with interfaces to the aircraft's FMS and tactical flight controls. AOP is able to leverage "intent" information from other traffic that may be obtained in the future through SWIM, ADS-B IN, or other means. Validating AOP's traffic separation algorithms embedded in TAP through years of operational use in the non-safety-critical TASAR and Strategic ABTM application of "user requests" will result in a robust and highly reliable trajectory planner that will have proven its integrity through a formal safety assessment process for the final step in the roadmap: Full ABTM.

Step Five - Full ABTM

Full ABTM, depicted in Figure 5, adds tactical separation assurance capability to the Strategic ABTM capability (i.e. downstream sector modifications to the route) such that autonomous trajectory changes can safely begin in the current sector rather than be delayed to the next sector. NASA has invested significant time and resources investigating this operation under the concept of "Autonomous Flight Rules" (AFR)[3] and determining the required functionality of onboard ABTM automation through development of the prototype "Autonomous Operations Planner" (AOP).[5] The ultimate objective of AFR and AOP is to enable operators to self-manage their trajectories in the NAS alongside conventionally

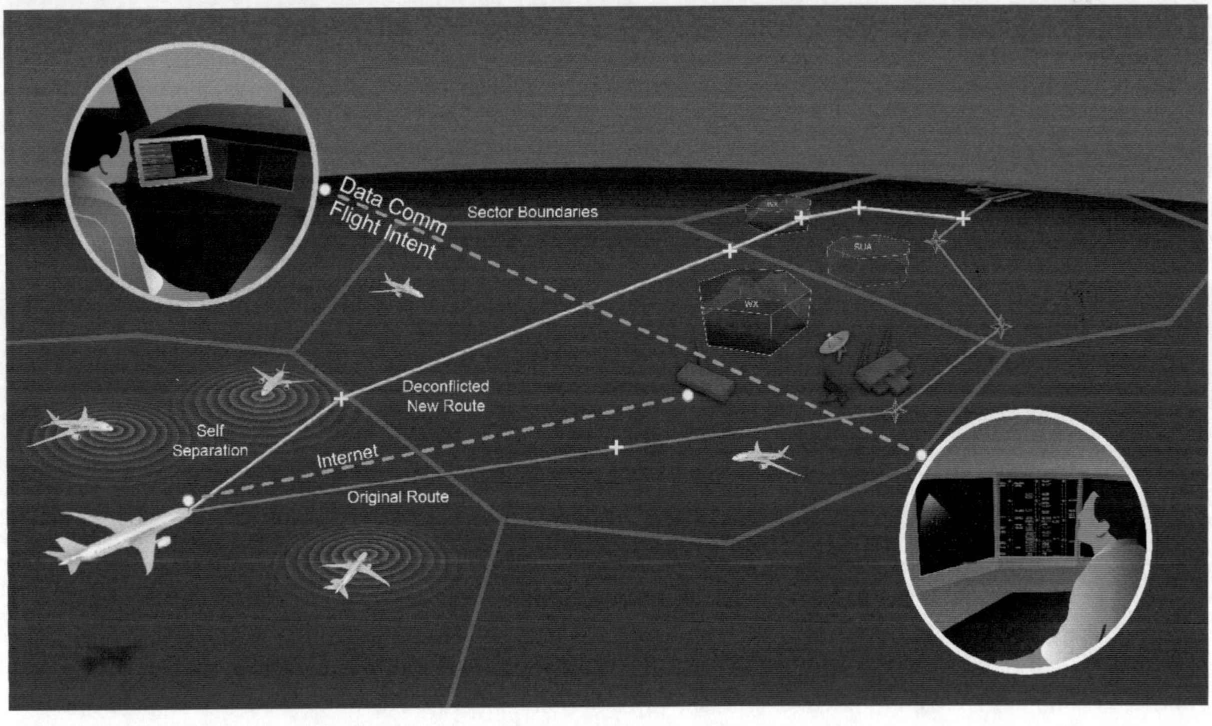

Figure 5 ABTM Roadmap Step 5 - Full ABTM.

managed traffic, preserving safety and fairness in the use of NAS resources but bypassing constraints.

AFR combines the flexibility of Visual Flight Rules (VFR) with the safety of Instrument Flight Rules (IFR). With the help of sensors and computational hardware, pilot displays provide full-time guidance along a continuously updated, wind-optimized trajectory, safely avoiding other traffic, active SUAs and weather hazards. AFR flights will participate in destination airport TBFM, but will employ optimized insertion into the traffic flow close to the landing runway. The benefits of flying the best wind route are very large and may be mostly achieved in earlier stages of ABTM, but these benefits are small in comparison to the delays resulting from TFM constraints in all its forms. In Full ABTM, the TAP system will be certified for the separation function. This equipment certification along with pilot training and operational approval will permit AFR operations for flights in the NAS at all altitudes and in all but active SUA airspace.

Achieving the approvals for AFR in FAR Part 121 operations may be time consuming but potentially accelerated by supporting experience gained using the concept first in Part 91 GA operations and in low altitude UAS operations. These flights also provide a logical environment to prove out reduced horizontal separation standards made possible using high update rate ADS-B IN that provides precise target positioning and short term intent. Reduced horizontal separation standards should be based, like collision avoidance systems, on the time to the closest approach and a minimal fixed protection area. Reducing the size of protected airspace in safe, legal separation would dramatically reduce the number of conflicts experienced in a given traffic density and the extent of the required resolution maneuvers.

The Way Forward

The FAA's NextGen programs are creating some essential technologies to modernize the way air traffic is managed. Augmented GPS for positioning, timing, and navigation in all phases of flight have made area navigation by all airspace users a reality. Instrument approaches may now be made to nearly every runway without expensive ground-based transmitters. Data communications is only just beginning to be realized, but it holds the promise of freeing aviation from the limitations of voice and extensive manual entry of flight information. The sharing of system status and weather data through SWIM is also in its infancy but can be used to help ensure safety by providing information at the right time and in all the right places. ADS-B provides the information needed to perform reliable traffic de-confliction and, ultimately, separation services on board the aircraft while sharing intent with all other NAS systems. But the lack of change in the control paradigms themselves has led to continued constraints imposed on IFR operations as a way of life, much to the frustration of system users. ABTM offers a way to overcome most of these constraints on properly equipped flights without impacting other traffic or conventional air traffic services.

In keeping with incremental improvements to TASAR capability and their resulting operating benefits to the first adopters, Step 1 is relatively simple, requiring only modest changes on the aircraft and no changes to ground systems or to pilot/controller procedures. It is essentially a tool used during the enroute cruise phase of flight. Step 2 increases the benefits to the first adopter by enabling longer and more complex trajectory changes for the entire flight outside the terminal airspace. The following steps continue to add benefit using logical extensions of NextGen and NASA developed tools. Though the greatest benefits to users will be achieved in the final step, Full ABTM, stopping at any step still adds value for the operator.

Most importantly, ABTM introduces these capabilities in a way that benefits the first adopter and every succeeding operator that chooses to equip. There is no requirement for a "critical mass" of users to equip or for ground systems to be modified, other than those changes currently planned for NextGen, to realize the benefits. Continued public/private partnering to refine and prove the benefits and safety of these new ABTM capabilities can ensure the initially envisioned transformation of ATM actually takes place.

References

1. Cotton, W., Hilb, R., Koczo, S. and Wing, D., *"A Vision and Roadmap for Increasing User Autonomy in Flight Operations in the National Airspace"*, 16th AIAA ATIO Conference, AIAA-2016-4212, 2016
2. Ballin, M., Wing, D., *"Traffic-Aware Strategic Aircrew Requests (TASAR)"*, 12th AIAA Aviation Technology, Integration, and Operations (ATIO) Conference, AIAA-2012-5623, 2012.
3. Wing, D. and Cotton, W.B., *"For Spacious Skies: Self-Separation with 'Autonomous Flight Rules' in US Domestic Airspace,"* 11th AIAA ATIO Conference, AIAA-2011-6865, 2011.
4. Woods S, Vivona R, Henderson J, Wing D, and Burke, K. "Traffic Aware Planner for Cockpit-based Trajectory Optimization." AIAA-2016-4067, 2016.
5. Karr D, Vivona R, Roscoe D, DePascale S, and Wing D. "Autonomous Operations Planner: A Flexible Platform for Research in Flight-Deck Support for Airborne Self-Separation." AIAA-2012-5417. 2012

Title:	**Driving Operations Monitoring and Decision Making by Leveraging Multidimensional Data Visualizations**

Author:	**Ankush P. Karnik**
Affiliation:	**IT WORKS! Inc.** **3416 9th Street N.E.** **Washington, DC 20017**
Project Affiliation:	**Federal Aviation Administration** **Air Traffic Organization** **Systems Operations** **Office of Performance Analysis** **1575 Eye St. NW** **Washington, DC 20005**
Corresponding Author:	**Ankush P. Karnik, MBA** **Project Manager** **IT WORKS! Inc.** **Email: ankush@itworksdc.com** **Tel: (202) 455 4781 x202**

Ankush P. Karnik, MBA

Abstract

Data collection and retrospective analysis are integral and important components of business operations in any industry, and particularly in aviation where data serves as a backbone of operations. To present findings that can lead to operational improvements, it is important that the synthesized data be presented in a readable, interpretable, and understandable format. Historically, traditional crosstabs/tables, bar charts, and line charts have been used for this purpose. Advancements in visualization tools now make it possible to present information in an interactive, multidimensional format which is easy to understand and interpret. Data visualization tools are enabling users to be more inquiring, and allowing them to delve further into the analysis and reasoning, thereby acting as a catalyst in driving decisions to improve operations. Optimal visualizations require thoughtful organization and preparation of the data. This paper focuses on the evolution of aviation visualizations within the Federal Aviation Administration's (FAA) Air Traffic Organization (ATO). It also describes how visualization tools are currently being adopted to help drive operational monitoring and excellence in the FAA ATO. Finally, lessons learned and best practices to develop meaningful and actionable multi-dimensional visualizations are mentioned.

Keywords: Digital Dashboards, Business Intelligence, Air Traffic Visualization, Operations Monitoring

Introduction

Data is everywhere! The Federal Aviation Administration (FAA) has been collecting and analyzing data related to flight delays for many years, and key performance indicators (KPIs) have been added over time to help monitor the performance of the National Airspace System (NAS), as well as a variety of reports and dashboards to help illustrate the trends in KPIs along with other necessary reportable information. These reports are used across all levels in the FAA to review operations, and make decisions to improve operation of the NAS.

Advancements in data architecture and storage strategy has allowed incorporating the large amount of new data and data sources to make them readily accessible and usable; however until recently the traditional two dimensional visualizations (bar charts, line graphs etc.) were still used to visualize the data, thereby, at times, limiting the ability to clearly represent complex situations in a concise manner. With two dimensional visualizations, multiple charts are often required to communicate the entire message, which dilutes the importance of the message, or obscures it altogether. Further, the FAA needed to employ computer programmers and specialized analysts to create such visualizations as many times it requires in-depth knowledge of databases and ETL (Extract, Transform, and Load) methods and tools, which front-line and operational personnel don't typically possess, thus limiting their ability to consume data for decision making purposes.

Recent advancements in data architecture and storage strategy have allowed incorporating new data sources to make them readily accessible, and advancements in data visualization and analysis tools have put the power of presentation and visualizations in the hands of the front line personnel, who may or may not have in-depth knowledge about databases or Structured Query Language (SQL). Creating complex visualizations is now easily achievable with little or no database knowledge, once the data has been structured and prepared for efficient retrieval and display. But with so much power in users' hands there are certainly best practices to be recommended to create meaningful visualizations that are actionable and cater to decision making in the organization.

This paper discusses complex visualizations developed to monitor operations and help in decision making process at the FAA to help enhance the safety and efficiency of air traffic. Specific use cases are provided illustrating the evolution of a single report through various time periods of visualization.

Growth of Data

KPIs and supplemental data from Technical Operations (TechOps), National Oceanic and Atmospheric Organization (NOAA), Mitre, Performance Data Analysis and Reporting System (PDARS), and National Traffic Management Log (NTML), and other FAA operational systems are used in analysis and reporting.

Circa 1997, when the FAA first began to make performance data readily available through the internet, monthly data feeds for three data sources were available – ETMS (today referred to as Traffic Flow Management System (TFMS)), Airline Service Quality Performance (ASQP), and Official Airline Guide (OAG) schedule data and fed into the Consolidated Operations and Delay Analysis System (CODAS).

Today, the FAA Office of Data Integration collects data from several data sources, such as NTML, TechOps, and NOAA, from which KPIs and supporting fields are extracted and fed into Aviation System Performance Metrics (ASPM) on a daily basis for reporting on many aspects of NAS performance. The graphic below shows contrast between 1997 and today, including the future where more data may be available on a more frequent basis for processing and visualizing.

The FAA has come a long way in terms of data collection, processing, storage, and aggregation, and additional improvements are coming down the road. This paper describes recent significant progress in the FAA's endeavor to make operational information for retrospective analysis and decision-making more readily available and customizable for front-line and support teams at the FAA.

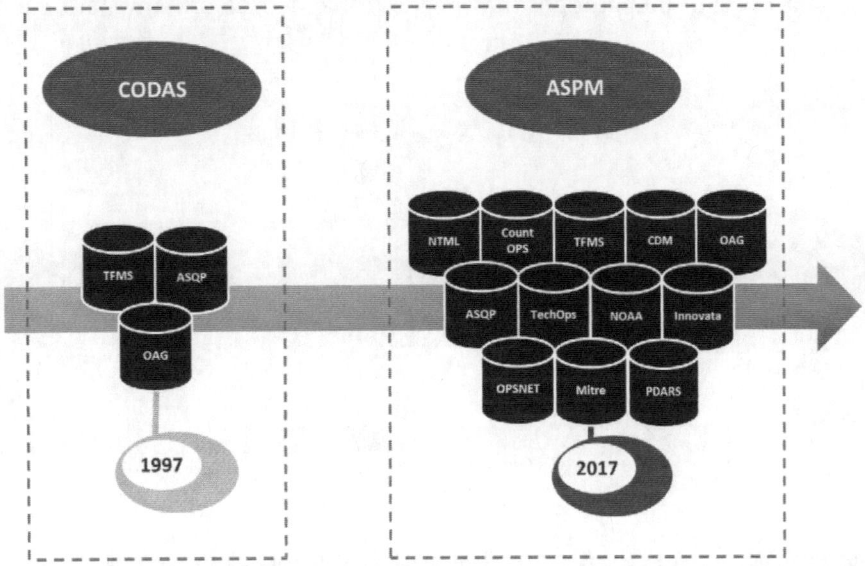

Figure 1: Growth of Data. Shown is growth of number of data sources from 1997 to 2017. CODAS: Consolidated Operations Delay Analysis System; TFMS: Traffic Flow Management System; ASQP: Airline Service Quality Performance; OAG: Official Airline Guide; ASPM: Aviation System Performance Metrics; NTML: National Traffic Management Log; CountOPS: FAA system utilizing data from National Offload Program (NOP); CDM: Collaborative Decision Making; TechOps: Technical Operations; Innovata: Data provider for flight schedule data; OPSNET: Operations Network; Mitre: Mitre Corporation; PDARS: Performance Data Analysis and Reporting System

Most of the data sources accessible to FAA analysts to date are structured. In other words, the data is highly organized within tables and databases, searchable via traditional search and query methods such as the SQL. Increasingly however, unstructured data is available to be pooled and analyzed along with structured data, and the combination can facilitate deeper actionable insights. Unstructured data may be difficult to store, and may not appear relevant or insightful when considered in isolation of the operating environment at the time, but may be immensely beneficial in deriving deep insights when considered and analyzed along with the structured data. An example of the benefits of unstructured data are assessing the outcome Traffic Management Initiatives (TMI) by supplementing information on operations, delays, and the TMIs duration and affected traffic, with customer comments providing their perspective on the impact. Additionally, collecting and structuring customer input over time can illuminate patterns and better allow the FAA to respond to challenging weather and traffic conditions, and also in planning for specific situations. The analysis could be further supplemented, and applied to other situations, by using radar data along with the conversations between Air Traffic Controller and the pilot to help provide essential information that may not be available today.

Data Preparation

Data Preparation is often an unspoken piece of the visualization puzzle. Today's business intelligence tools make it very easy for users to generate visualizations by connecting to the data sources and dragging or dropping fields onto columns and rows shelves. The unspoken truth, however, is that much; if not most data sources require significant preparation before they can be ingested by business intelligence tools for meaningful and accurate displays. As the graphic below shows, there is an important component of 'data preparation' that needs to happen before any data can be considered visualization ready.

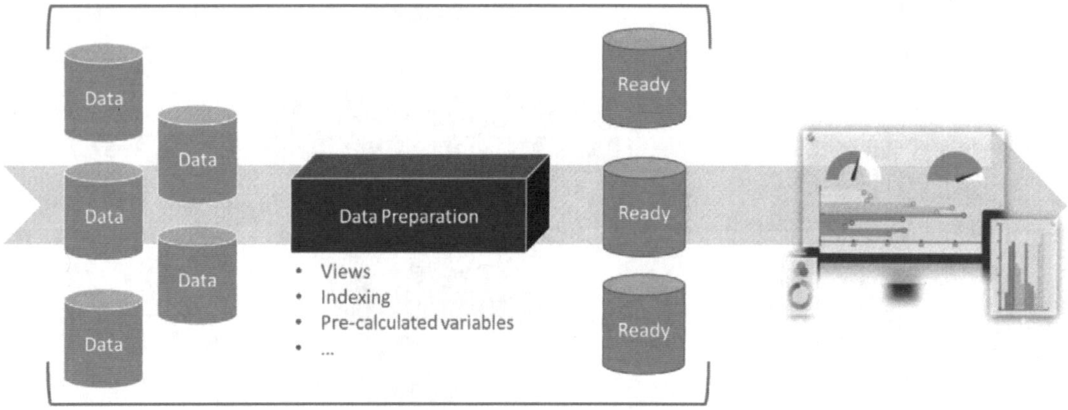

Figure 2: Data preparation[1]: Data preparation is an important component to make data visualization ready.

Preparation can involve a variety of steps but generally includes the following:

- **Indexing:** Indexing is defined as applying structure to the data that helps improve speed of data retrieval and operations on the database table. [1] It is best practice to index tables utilized for visualizations.

- **Pre-Calculating Variables**: Business intelligence tools are fully capable of performing calculations on the fly while generating visualizations. If the calculation is complex, however, it is best practice to pre-calculate values, store them in the table/view, and then make them field available to the visualization. This is because the resources needed for a complex calculation may be significant and deter performance of the dashboards/ visualizations in the long run.

- **Database views:** Database view is a searchable object in a database that is defined by a query[2]. A view does not store data; it consolidates data from various tables into a single virtual table. As

[1] Karnik, Ankush, "Tableau Brown Bag" (unpublished presentation, February 28, 2017),PDF and PowerPoint files.

dashboards become more complex and the need for full drill down capability gets established, it is always a good strategy to create views from the original tables and store them in a separate schema ready for access by a business intelligence tool. Views can be set to update regularly to ensure recent data is available for display on the dashboard.

The above mentioned data preparations steps were key in generating visualizations mentioned in this paper. If skipped, the constructed dashboards would have been difficult to use particularly due to performance related issues. It is worth noting that each of these dashboards is pulling well over a million records at a time. Some of the tables are large both length and width-wise, and therefore data preparation is an extremely important and necessary step in the dashboard construction process.

Evolutions of Air Traffic Dashboards

Early visualizations of air traffic data that pre-date Aviation System Performance Metrics (ASPM) systems were sophisticated at the time, but basic compared to current visualizations. The evolution of computers, data storage, networks, and software have made it possible to quickly process and synthesize increasing amounts of data in support analysis and decision-making.

NAS Daily Report

One of the first dashboards used in in daily senior management briefings at the FAA was the NAS Daily. It is a static, tabular report without graphics or drill-down capabilities. The FAA end-user cannot manipulate or augment the data contained in the report, and any changes requested by the FAA must be implemented by a programmer.

This report enables users to conduct some secondary analyses including comparison of two different time periods, but with no drill down capabilities.

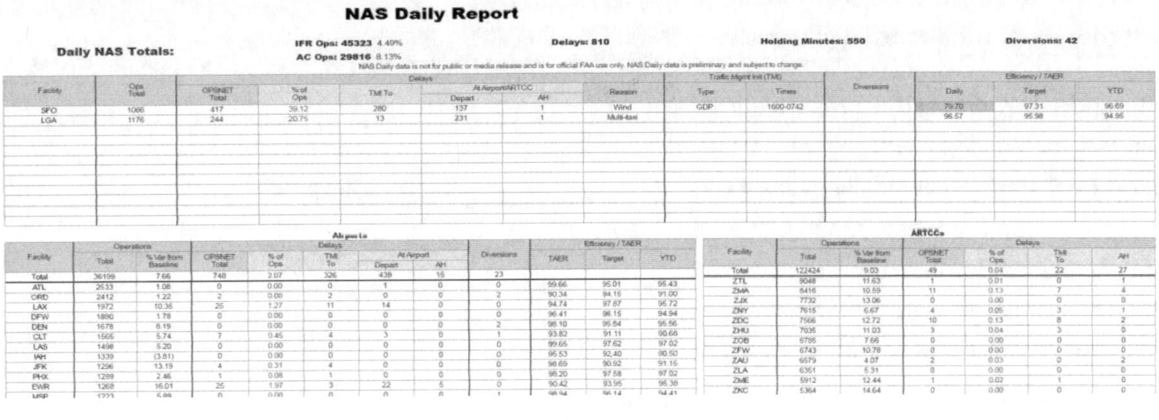

NAS Daily Report

Daily NAS Totals: **IFR Ops: 45323** 4.49% **AC Ops: 29816** 8.13% **Delays: 810** **Holding Minutes: 550** **Diversions: 42**

NAS Daily data is not for public or media release and is for official FAA use only. NAS Daily data is preliminary and subject to change.

Facility	Ops Total	Delays						Traffic Mgmt Init (TMI)		Diversions	Efficiency / TAER		
		OPSNET Total	% of Ops	TMI To	At Airport/ARTCC Depart	At Airport/ARTCC AH	Reason	Type	Times		Daily	Target	YTD
SFO	1066	417	39.12	280	137	1	Wind	GDP	1600-0742		79.70	97.31	96.69
LGA	1176	244	20.75	13	231	1	Multi-taxi				96.57	95.98	94.95

Airports

Facility	Operations		Delays					Diversions	Efficiency / TAER		
	Total	% Var from Baseline	OPSNET Total	% of Ops	TMI To	At Airport Depart	At Airport AH		TAER	Target	YTD
Total	36199	7.66	748	2.07	326	439	15	23			
ATL	2533	1.08	0	0.00	0	1	0	0	99.66	95.01	95.43
ORD	2412	1.22	2	0.08	2	0	0	2	90.34	94.15	91.00
LAX	1972	10.36	25	1.27	11	14	0	0	94.74	97.87	95.72
DFW	1890	1.78	0	0.00	0	0	0	0	96.41	98.15	94.94
DEN	1678	8.19	0	0.00	0	0	0	2	98.10	95.64	95.56
CLT	1565	5.74	7	0.45	4	3	0	1	93.82	91.11	90.68
LAS	1498	5.20	0	0.00	0	0	0	0	99.65	97.62	97.02
IAH	1339	(3.81)	0	0.00	0	0	0	0	95.53	92.40	90.50
JFK	1296	13.19	4	0.31	4	0	0	0	98.69	90.92	91.15
PHX	1289	2.46	1	0.08	1	0	0	0	98.20	97.58	97.02
EWR	1268	16.01	25	1.97	3	22	5	0	90.42	93.95	95.38
MSP	1223	[illegible]	0	0.00	0	0	0	1	98.94	96.14	94.41

ARTCCs

Facility	Operations		Delays			
	Total	% Var from Baseline	OPSNET Total	% of Ops	TMI To	AH
Total	122424	9.03	49	0.04	22	27
ZTL	9048	11.63	1	0.01	0	1
ZMA	8416	10.59	11	0.13	7	4
ZJX	7732	13.06	0	0.00	0	0
ZNY	7615	6.67	4	0.05	3	1
ZDC	7566	12.72	10	0.13	8	2
ZHU	7035	11.03	3	0.04	3	0
ZOB	6785	7.66	0	0.00	0	0
ZFW	6743	10.78	0	0.00	0	0
ZAU	6579	4.07	2	0.03	0	2
ZLA	6351	6.31	0	0.00	0	0
ZME	5912	12.44	1	0.02	1	0
ZKC	5364	14.64	0	0.00	0	0

Figure 3: NAS Daily Report: Example of NAS Daily report showing key metrics/KPIs for Core airports.

[2] "Essential SQL", accessed April 3, 2017, https://www.essentialsql.com/what-is-a-relational-database-view/.

Air Traffic Organization (ATO) Efficiency Report Online (AERO)

Since implementing ASPM, the visualizations, even though in a very simplistic form have helped in understanding the aviation data across all levels of management in the Federal Aviation Administration (FAA) and monitoring operations. Air Traffic Organization (ATO) Efficiency Report Online or AERO provides both a summary and a comprehensive view of the day for a specific facility[3]. While the format of this report is simple and easy to understand, it is generated from a complex network of data sources and aggregation, and which required several programming hours to generate the report.

For each selected location, this report provides a summary view of Operations, Weather, Efficiency Metrics such as Terminal Arrival Efficiency Rate (TAER), System Arrival Efficiency Report (SAER), Taxi-in and Taxi-out, Go Arounds, and Equipment Availability. An hourly trend is shown for Taxi-in, Taxi-out, TAER, Arrivals, and Departures. Additionally 90-day trends are provided for Taxi-in, Taxi-out, Go Arounds, and TAER. In order to construct this report, data is aggregated from multiple data sources such as Operations Network (OPSNET), ASPM, COUNTOPS, and PDARS, queries are developed to extract data from each of these data sources and aggregate it to a desired level.

The AERO report was conceptualized and designed by the need to visualize ATO data and metrics and allow for easy understanding of 'the day' at a given facility, and also aid in decision making based on the operational performance at each facility. Drill down capabilities are provided, where the user can click on any of the metrics provided on the report and drill down to the next level of detail, in most cases to the individual flight level.

This report has enabled a range of users from senior management to Air Traffic operational staff to get a daily overview as well focus in on the most granular level of data. A related dashboard called the NAS AERO consolidates information on thirty core airports into one report for a more system-wide view.

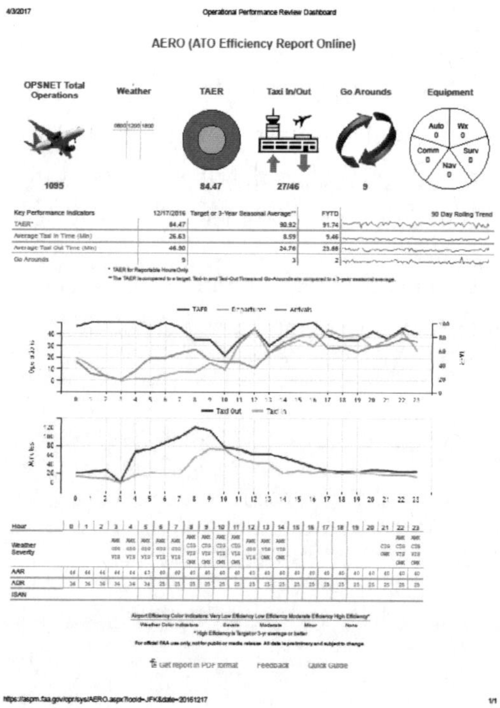

Figure 4: ATO Efficiency Report Online (AERO): Example AERO report for selected airport and date.

[3] "Aviation System Performance Metrics (ASPM)", accessed April 7, 2017, aspm.faa.gov.

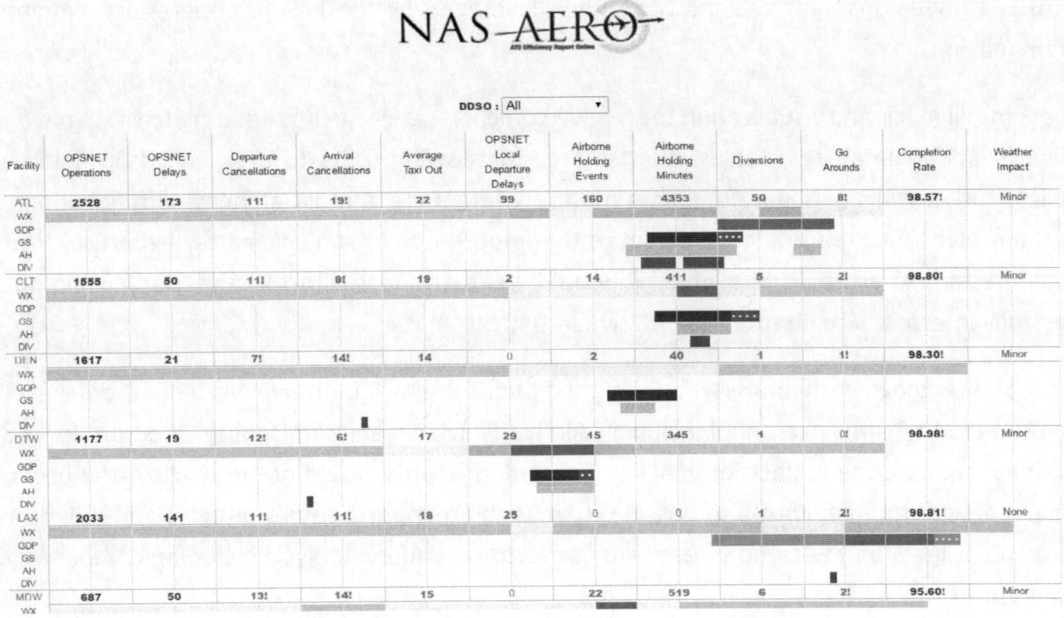

Figure 5: NAS AERO Report: Example of NAS AERO report showing metrics/KPIs for core airports.

Both the AERO Report and the NAS AERO report provide critical information to the users, and both these reports are actively used by all levels of management in daily FAA ATO operations monitoring and decision making. However, these reports still fall short, not because they lack information, but because the visualizations used for these reports limit the features, and it would be inherently complex and busy if both the reports would need to be combined. Additionally, both these reports are prepared to be print-ready vs. screen-ready, thereby preventing users from interacting with these reports.

Both these reports and many others like these provide a good summary of the operations, but also create a need for a secondary analysis, which is a more in-depth analysis of KPIs at each facility. Typically a secondary analysis entails drilling down in the data, conducting analysis, and visualizing the result. Also traditionally, secondary analysis reports / charts have been kept independent than the summary reports / charts as it has been difficult to combine, or a combination thereof results in information overload for the consumer.

In order to maintain clarity and impact of the information, it must be presented in a clear and concise manner. Interactivity also enables users to be more engaged and inquiring about the information. It also helps them understand the interconnectivity and relationships between KPIs and other metrics.

Visualizing Air Traffic Data – Interactive Dashboards

To that end, in 2016 FAA Office of Performance Analysis launched an initiative to leverage a business intelligence platform and transform static PDF reports to interactive dashboards that can be accessed by range of users within the organization, and the users can drill down to the most granular level of the data that is available, typically the flight level. Next section provides examples of dashboards constructed under this initiative and have been leveraged by the FAA for operations monitoring and decision making.

Business intelligence and visualization tools have come a long way from requiring users to have technical knowledge about databases, tables, queries etc., to allowing users to drag and drop fields to create impactful and engaging visualizations in a matter of minutes. Almost all heavy lifting is done by such tools themselves. But this does not minimize the importance of subject matter expertise. Without the subject matter expertise in the area, it would be extremely difficult and perhaps impossible to successfully interpret visualizations and derive insights from it.

First set of interactive dashboards to be designed under the new initiative were operational dashboards to be used by Air Traffic personnel including Air Traffic Managers and Deputy Directors for monitoring operations and assisting in decision-making. These dashboards would not only allow the users to get a summary view of the data, but also provide them with an opportunity to interact and drill down to the most granular level on the same screen, and perform secondary analysis including comparing different time periods.

FAA defines secondary analysis as a more in-depth research and analysis resulting from a summary review of NAS performance for the prior day. The result of secondary analysis is discussing findings which serve as a major input into executive and operational decision making required to efficiently operate the NAS.

Below are some examples of interactive dashboards developed and deployed to production at the FAA, and used by operational staff on a routine basis.

Facility Summary Dashboard

Facility summary dashboard is positioned as an enhancement to the current Daily AERO and the NAS AERO reports. It provides the same information as on the Daily AERO report but in a different representation including comparison of current day metrics to established benchmarks.

Below figure shows the Facility Summary dashboard for JFK airport for a specific day. On this day, there was a weather event which affected operations and efficiency metrics at the airport. The top section of the dashboard provides summary metrics for the selected day for JFK airport and the bottom section provides hourly view of the metrics starting at 0000 hours through 2300 hours.

This dashboard is extremely helpful for Facility Managers and Air Traffic Specialists to take a 'day's' view at the facility. They can also compare KPIs on 'like' days to determine any differences or conduct secondary research.

Dashboard Construction

To assemble the KPIs on the dashboard the data is pulled from 5 different data sources which are essentially views built from the same data that is available in ASPM.

The summary section of dashboard shows KPI values for the selected day. The color coding in the top section is based on comparison to established benchmarks. For example – red indicates that the current value is over established benchmark and green indicates that the current value is under the established benchmark.

In the chart section of the dashboard, details for a selected day are shown by the hour. Various KPIs are shown including, Airport Arrival Rate (AAR), Airport Departure Rate (ADR), Average Taxi In, Average Taxi Out, Actual Departures, Actual Arrivals, and Terminal Arrival Efficiency Rate (TAER). This dashboard also overlays weather impact (None, Minor, Moderate, Severe) to add more meaning to the fluctuation in KPIs throughout the day.

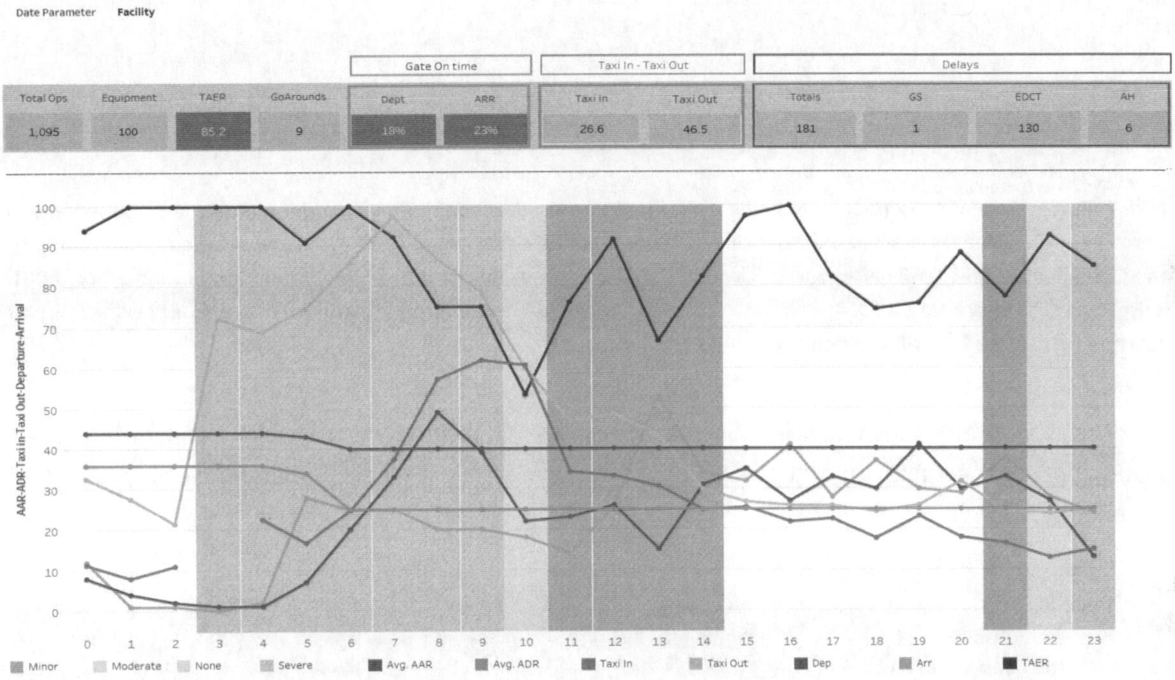

Figure 6: Facility Summary Dashboard: Example of Facility Summary Dashboard showing daily summary and hourly trend of metrics/KPIs for selected airport and date. Total Ops: Total Operations; TAER: Terminal Arrival Efficiency Rate; Dept / Dep: Departures; ARR: Arrivals; GS: Ground Stops; EDCT: Expect Departure Clearance Times; AH: Airborne Holding; Avg. AAR: Average Airport Arrival Rate; Avg. ADR: Average Airport Departure Rate; None-Minor-Moderate-Severe: Weather impacts

A pop up feature, displayed below, displays the runway configuration and METAR for a selected hour. When an hour is selected, all the other hours fade, to help the user to focus on the time period in question.

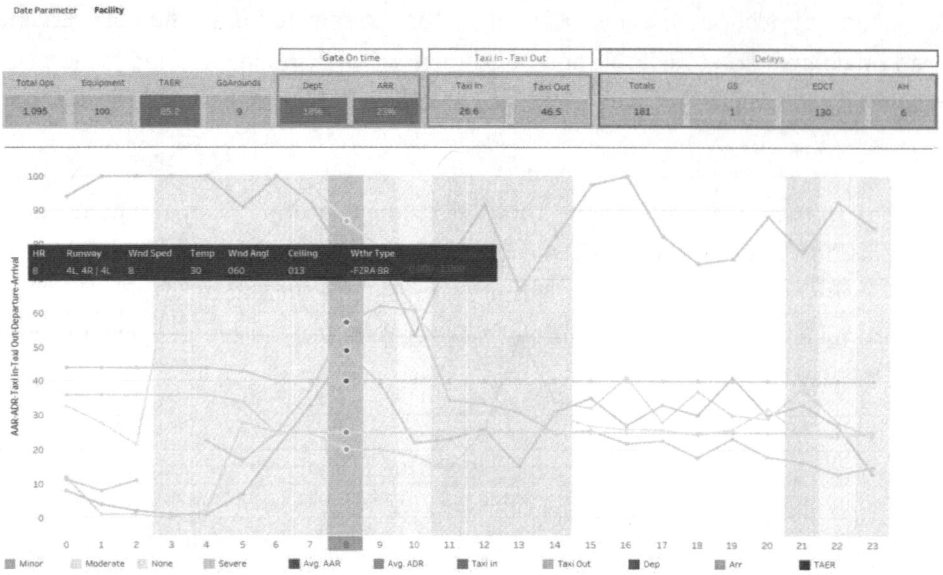

Figure 7: Facility Summary Dashboard: Example of Facility Summary Dashboard showing daily summary and hourly trend of metrics/KPIs for selected airport and date, including runway and weather details overlay for the selected hour. Total Ops: Total Operations; TAER: Terminal Arrival Efficiency Rate; Dept / Dep: Departures; ARR: Arrivals; GS: Ground Stops; EDCT: Expect Departure Clearance Times; AH: Airborne Holding; Avg. AAR: Average Airport Arrival Rate; Avg. ADR: Average Airport Departure Rate; None-Minor-Moderate-Severe: Weather impacts

The fade feature is also activated when one particular KPI is selected, which enables the user to focus on that KPI, but at the same time keep the other metrics visible to help understand the relationship between KPIs.

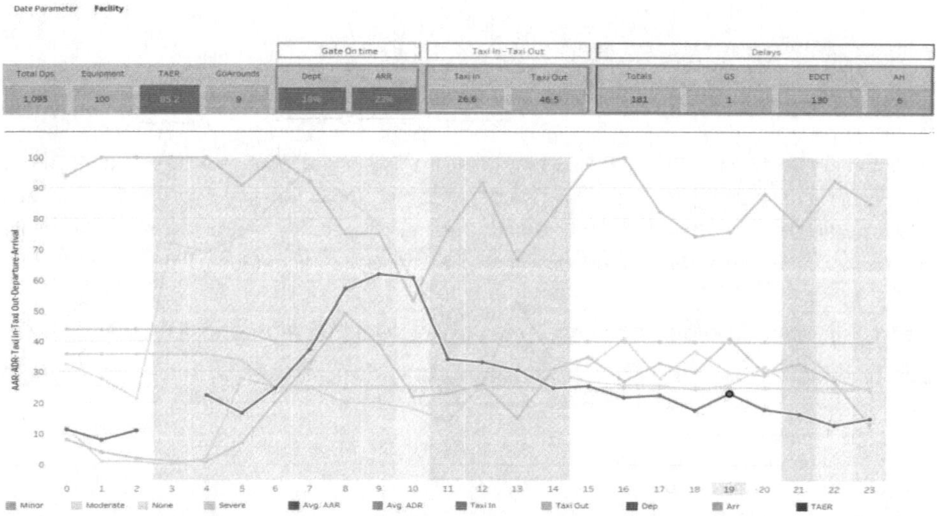

Figure 8: Facility Summary Dashboard: Example of Facility Summary Dashboard showing daily summary and hourly trend of metrics/KPIs for selected airport and date with Taxi-In trend highlighted. Total Ops: Total Operations; TAER: Terminal Arrival Efficiency Rate; Dept / Dep: Departures; ARR: Arrivals; GS: Ground Stops; EDCT: Expect Departure Clearance Times; AH: Airborne Holding; Avg. AAR: Average Airport Arrival Rate; Avg. ADR: Average Airport Departure Rate; None-Minor-Moderate-Severe: Weather impacts

Impact on Operations and Performance Monitoring

In addition to providing facility managers with a summary of a day's performance, the Facility Summary Dashboard can also be used to compare performance across days. For example at a recent briefing to the executives at the FAA, a comparison between two days with similar weather events for a single facility triggered discussions on best practices and approaches to recover operations and efficiency after a weather event.

Another significant benefit provided by this dashboard is consolidation of valuable information on a single screen. Prior to the Facility Summary dashboard, users had to refer to multiple static reports in order to provide a comprehensive picture, and this process was extremely time consuming and slow.

Implementing the pop up for METAR and runway configuration information has saved users time and are now able to pull all necessary information for the briefing from one single dashboard.

Future enhancements include addition of METAR and SPECI information to this dashboard, which will provide more relevant and complete information to operational staff. Dashboards such as this one helps Air Traffic personnel visualize data and obtain critical information that helps in decision making to efficiently operate the National Airspace System (NAS).

Looking at the above figures, the Facility Summary dashboard paints a comprehensive picture of the day at the JFK airport. From the dashboard it can be seen that the severe weather impacted arrivals and departures at the airport. There was capacity, but Air Traffic Control could not get flights in or out of the facility. As the weather eased, departures can be seen picking up and at the same time taxi times increase, this is mainly as the aircrafts line up on the taxi way for departures. Between 1500 hours and 2100 hours we can see that there was no weather, and arrivals and departures were within the expected capacity, which is generally an example of a normal operation.

The Facility Summary dashboard is a good example an actionable dashboard that can be used to derive insights and drive operational decisions.

Operations Drill Down Dashboard

Operations Drill Down dashboard provides a summary of operations at Core 30 airports in the NAS for any given day. Users are able to select a Region and Date they want to review the data for.

Dashboard Construction

Information on this dashboard flows from left to right and top to bottom. The top left chart in this dashboard provides a summary view of the number of operations across all Centers, TRACONS, and Airports within the selected region for a specific date. The lower section of the dashboard provides dissection of the operations along various dimensions including the Type of Operations, Arrivals and Departures, and critical KPIs for ATO such as Completion Rate, Average Daily Capacity (ADC) and TAER.

Similar to the Facility Summary dashboard, multiple data sources are utilized to provide users with full drill down capabilities up to the Air Carrier level.

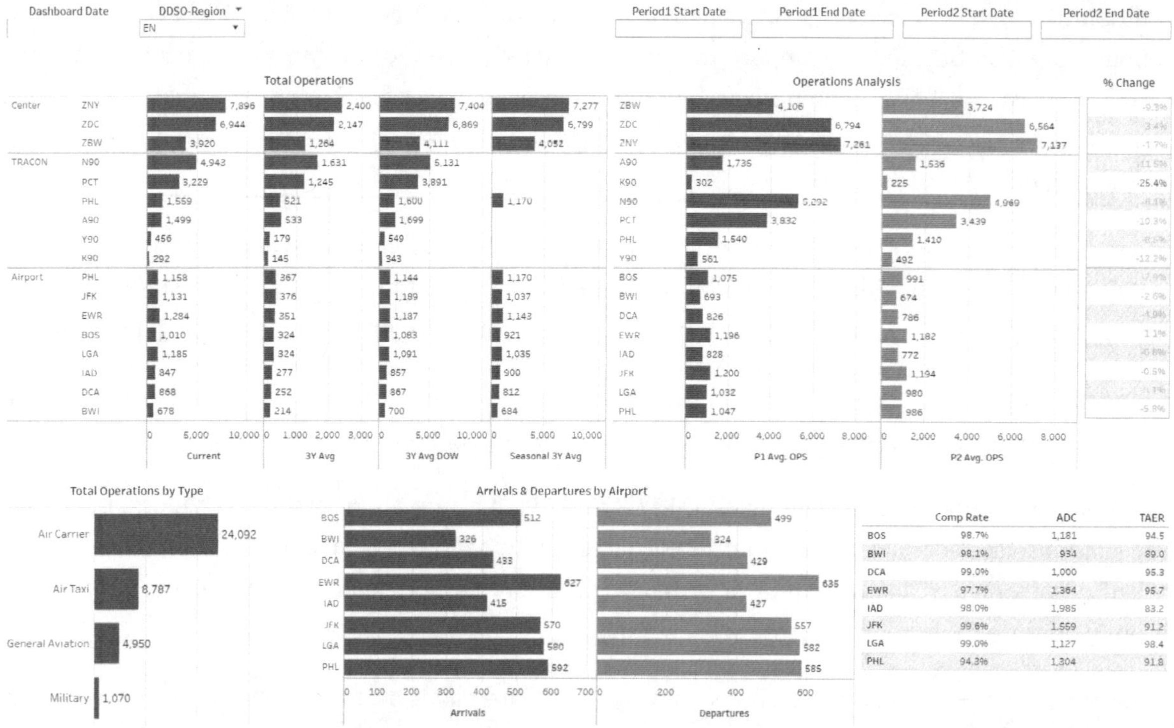

Figure 9: Operations Drill Down dashboard: Example of Operations Drill Down dashboard with full drill down capabilities. Summary and detailed metrics/KPIs shown for selected date and region. Comparisons shown for selected dates. Current: Metrics for selected date; 3Y Avg: 3 year average; 3Y Avg DOW: 3 year average for day of week; Seasonal 3Y Avg: Seasonal 3 year average; P1 Avg Ops: Average operations for period 1; P2 Avg OPS: Average operations for period 2; Comp Rate: Completion Rate; ADC: ADC: Average Daily Capacity; TAER: Terminal Arrival Efficiency Rate

Furthermore, users also have the ability to select on screen any group of airports or facilities they wish to analyze, and the dashboard will refresh all charts based on the selection. Figure below shows metrics from NY area airports which is driven by selecting EWR, JFK, and LGA in the top left chart.

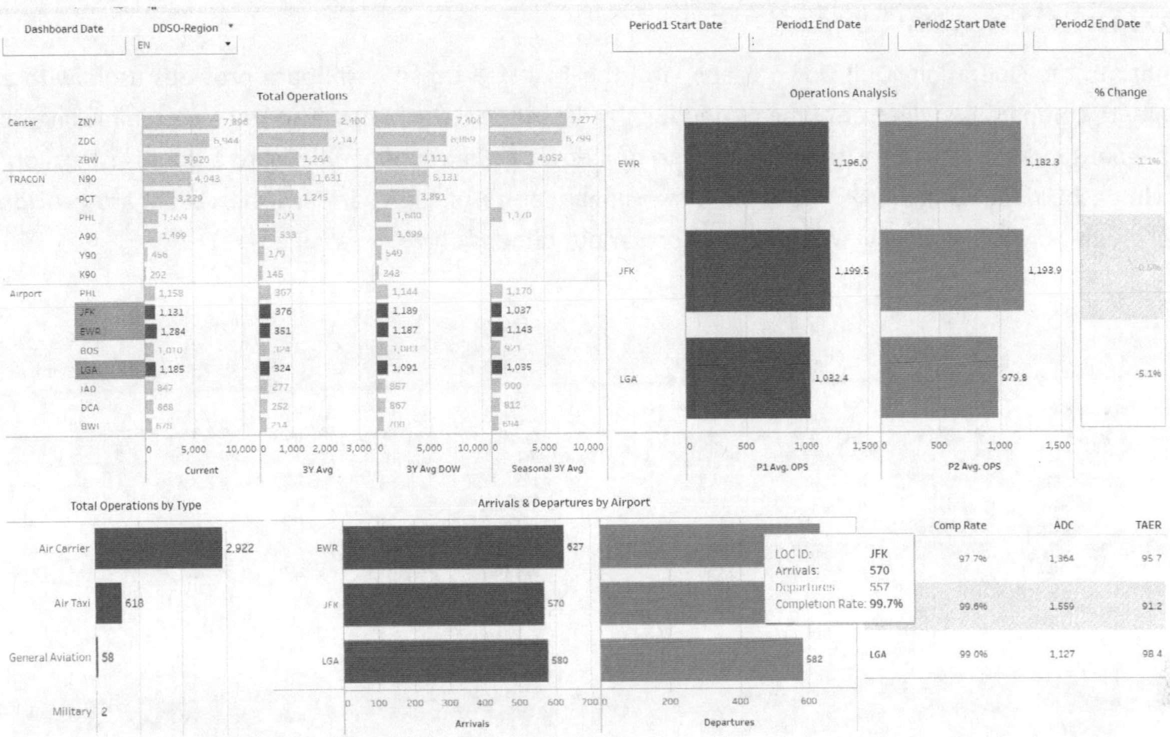

Figure 10: Operations Drill Down dashboard: Example of Operations Drill Down dashboard with New York area airports selected. Comparisons shown for selected dates and airports. Current: Metrics for selected date; 3Y Avg: 3 year average; 3Y Avg DOW: 3 year average for day of week; Seasonal 3Y Avg: Seasonal 3 year average; P1 Avg Ops: Average operations for period 1; P2 Avg OPS: Average operations for period 2; Comp Rate: Completion Rate; ADC: ADC: Average Daily Capacity; TAER: Terminal Arrival Efficiency Rate

Impact on Operations and Performance Monitoring

This dashboard is designed for Facility Managers, Director and Deputy Directors of Systems Operations, and all Air Traffic personnel to monitor operations within the region. Since this dashboard groups facilities by type – Airports, TRACONS, and Centers, it paints a good picture of the system load at a specific facility/facilities or region at a given time.

This dashboard is also an invaluable tool while planning for major events such as Super Bowl or other major sporting events, which creates a spike in aviation activity within the region, resulting in increase in operations managed by the Center, TRACON, and airports in the region. Increasing operations mean increase in demand and potentially capacity, which requires additional staff to efficiently operate the NAS. As an example, Operations Drill Down dashboard can serve as an input to the planners while planning for Super Bowl 2018 in Minneapolis, MN.

Future enhancements to this dashboard include integration of specific airports to help in monitoring and planning for special events, and inclusion of operations by air carrier for the selection region and facility.

Taxi Drill Down Dashboard

Similar to the Operations Drill Down dashboard, the Taxi Drill Down dashboard provides users with an ability to compare two different time period for Taxi In and Taxi Out times. The key factor in making this dashboard more actionable is to allow users to drill down from a summary level to a granular level on a single dashboard. Additionally, the multi-dimensional nature of the charts on the dashboard provides users with a cross-sectional view of the data on various dimensions.

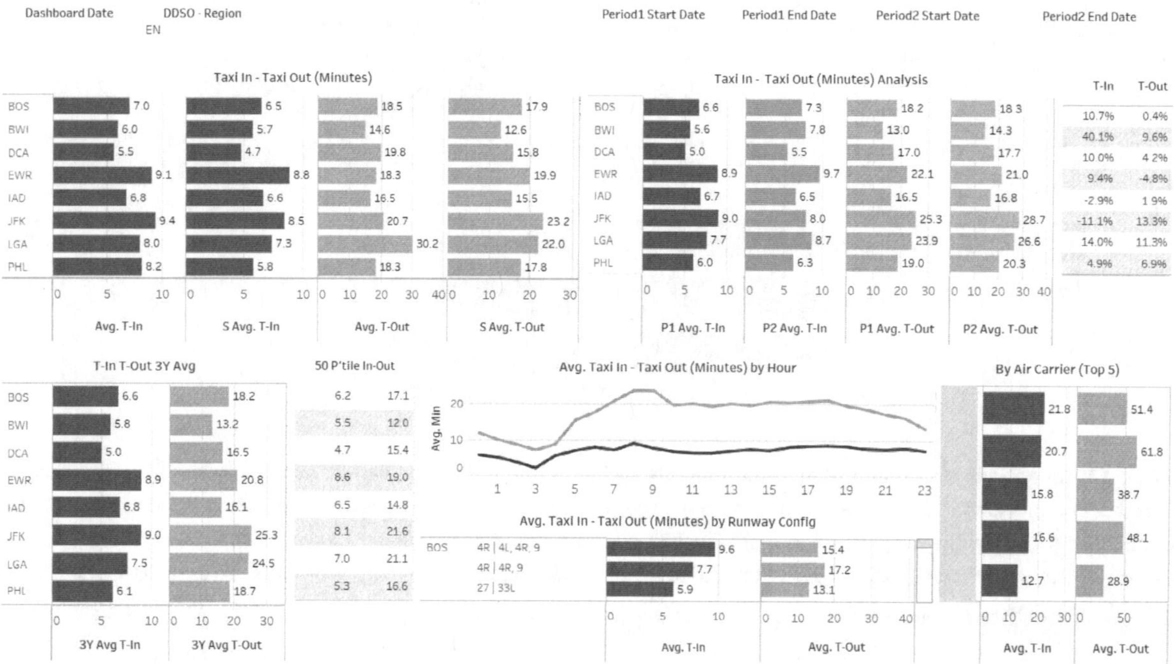

Figure 11: Taxi Drill Down dashboard: Example of Taxi Drill Down dashboard with full drill down capabilities. Comparisons shown for selected dates. Avg T-In: Average Taxi In; S. Avg T-in: Seasonal Average Tax In; Avg T-Out: Average Taxi Out; S. Avg T-Out: Seasonal Average Taxi Out; P1 Avg T-In: Average Tax In Period 1; P2 Avg T-In: Average Taxi In Period 2; ; P1 Avg T-Out: Average Tax Out Period 1; P2 Avg T-Out: Average Taxi Out Period 2; T-In: Percent change Tax in; T-Out: Percent change Taxi Out; 3Y Avg T-In: 3 year average Taxi In; 3Y Avg T-Out: 3 year average Taxi Out

Dashboard Construction

For example, the top left section of the dashboard lists summary metrics for Taxi In and Taxi Out. Here selected day's values are compared to Seasonal averages, which is an established comparative benchmark for the FAA. The top right section of the dashboard provides users an ability to compare two different time frames. However, the key categorizing this dashboard as actionable vs. informative dashboard are the charts in the lower half section of the dashboard.

These charts aggregate data along multiple dimensions providing a multi-dimensional view on a single screen. For example, we can see Taxi Times by the hour of the day, by runway configuration, and by air carrier. The users can also select a specific airport from the summary chart, which then filters the dashboard for that specific airport. User can also select a specific metric i.e. Taxi In or Taxi Out at a specific hour in the line chart, and the entire dashboard will filter and provide the runway configuration in effect at the time, and the airlines taxiing at that hour as shown in the figure below.

Impact on Operations and Performance Monitoring

Similar to other dashboards mentioned in this paper, Taxi Drill Down dashboard is designed for Facility Managers, Director and Deputy Directors of Systems Operations, and all Air Traffic personnel to monitor Taxi times within a region and for a specific facility.

Taxi time is a better indicator of Runway Usage, and ADC, which is a critical metric for individual facilities. Higher taxi times lead to delays, which then leads to reportable delays if they are beyond 15 minutes. Air Traffic managers and other personnel pay a close attention to all reportable delays.

Taxi Drill Down dashboard provides an excellent visual into the Taxi data for a region and a specific facility. It also allows user to compare two different time periods to evaluate the day's Taxi In and Out metrics to any period of time and to the established benchmarks.

Additionally, the dashboard also shows the runway configuration in use at a specific hour and corresponding Taxi In and Taxi Out time including the Air Carriers that were Taxiing at the selected hour. Such visualizations have helped Air Traffic personnel respond to queries such as:

- What was the loading on specific runway?
- Did higher taxi times translate into reportable delays?
- Why wasn't a specific runway used at a specific time when we were observing departure delays?
- Which carrier(s) or a specific flight(s) was a driver of taxi times at a given time?

Future enhancements to this dashboard include providing taxi information by specific runway by air carrier for each facility within the selected region and integrating with the Delay dashboards to help correlate reportable delays to high Taxi Out time at a specific facility for a specific day.

Go Arounds Drill Down Dashboard (Map Integration)

The Go Arounds Drill Down dashboard leverages the tool's ability to provide multi-dimensional representation of the data and also provide users with an ability to simulate a particular day / time.

Dashboard Construction

Shown below is a dashboard that illustrates Go-Around information at JFK airport. This information is available for all airports with ASPM data available. The upper half of the dashboard illustrates detailed flight level data for Go Arounds including Altitude, Range from Airport, Return Distance from Airport, and Return Time to Airport.

The map is generated using the Latitude and Longitude of the position where the event occurred according to PDARS data source. Also represented is the time of the day when the event occurred. Traditionally, this would have been another column in the chart, but instead is represented as a color.

By incorporating visuals such as a map, and representing data in the format familiar to the Air Traffic personnel, we are able to convey large about of information in clear and succinct manner without labeling each variables/measure shown on the dashboard. For example, since the map represents Go Arounds, and the labeling is done as typically seen on an ATC radar screen, we don't need to call out each element shown in the map.

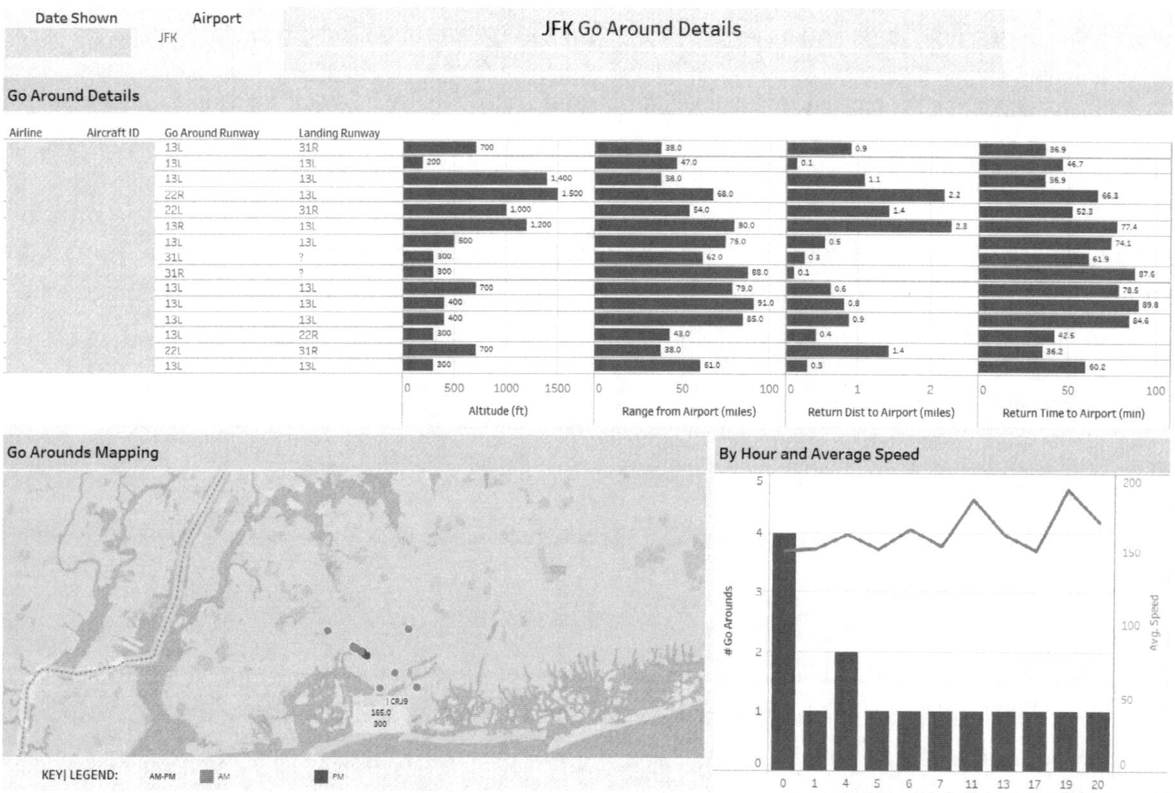

Figure 12: Go Arounds Drill Down dashboard: Example of Go Arounds Down dashboard with drill down capabilities and map integration.

Shown on the map is the period of the day when the Go Around event occurred, Air Frame number, Aircraft Type, Speed, and Altitude.

Impact on Operations and Performance Monitoring

The Go Arounds Drill Down dashboard is designed for Facility Managers, Director and Deputy Directors of Systems Operations, and all Air Traffic personnel at a particular facility. This dashboard is a spinoff of the main Go Arounds Drill Down dashboard and is developed to monitor Go Arounds at the New York area airports.

With the map representation the users are able to see the location of the event. Additionally selection of multiple facilities on the dashboard enables users to see Go Arounds in a specific region. Prior to the Go Arounds Drill Down dashboard, the users had to look at multiple reports to obtain required data, which was time consuming.

Potential enhancements to this dashboard include considering weather and other data that will help explain the cause of the specific Go Around.

Many other dashboards have been developed as part of the FAA ATO initiative led and managed by the Office of Performance Analysis, which collectively allows users to interact with the operational data and derive actionable insights. The tool also allows users to export the dashboard information in multiple formats including JPG, PDF, XLS, for further analysis and distribution.

The Adoption Path

All operational dashboards developed under this initiative have been in production for less than six months. Even though the dashboards are in their infancy stage, they are being adopted quickly throughout the organization. Air Traffic Specialists from various regions are accessing these dashboards on a regular basis to conduct the secondary analysis and report to operational managers and executives on a regular basis.

A steady rise in traffic to the dashboards is reported, with expectation that the traffic will increase multi-fold as the dashboards mature and are enhanced with additional features, meaningful visualizations, and more data to tell a story.

In order to facilitate agency wide adoption of these dashboards an information session is planned, and a SharePoint site is established for users to provide comments and feedback.

While most dashboards are currently open to internal FAA users, there are specialty use dashboards intended to be used by both internal FAA users and external stakeholders including air carriers. An example of one such dashboard is a runway usage and summary dashboard for the New York area airports developed in response to a construction project at JFK airport.

Users have also made requests for additional and more complex dashboards that combine different KPIs for various time periods to provide a causal – impact analysis, a key component of the secondary analysis and research.

Lessons Learned

The dashboard development team documented several lessons learned in the process of developing and implementing the dashboard. Some key lessons are highlighted below.

- **Data preparation is a key to building a successful dashboard**: Ensure that the data is available and visualization ready. Having access to the data does not mean it is ready to be visualized. Data must be cleaned, normalized, and stored in an accessible location for visualization.

- **Manage user expectations**: While managing user expectations is not new to software or application development processes, it is more important in case of development of digital dashboards. Complex visualizations can be generate in a short amount of time, and can open up possibilities to show big picture operation, but at the same time it can also be confusing to the users especially if they are used to looking at bar charts, tables, and other familiar visualizations. It is therefore important to know and understand the user base and slowly introduce them to newer – more elaborate visualizations so they understand the incremental value of them.

- **Digital dashboard is not a replacement for a report:** Business intelligence tools can be used to develop both reports and dashboards. However there is a difference between the two. A report depicts KPIs at a point in time and seldom allows user to drill down into the data. A dashboard, on the other hand, is used more for research and analysis. Therefore, while it tempting to convert all Excel and PowerPoint reports to dashboard, a question must be asked on how can these reports be made more interactive and what additional information can be provided so that the users can harness the full power of a dashboard developed using a business intelligence tool.

- **Beware of limited real-estate:** Dashboards are generally expected to be comprehensive and impactful. As such it is tempting to include a lot of information on a dashboard. However, attention must be paid to the amount of information shown on a single dashboard as more information generally means confusion.

- **Beware of varied aggregation levels:** Varied aggregation levels of data create a problem for a dashboard that is expected to have a full drill down capability. For example, if it is expected that a dashboard be developed to provide a daily level summary for a specific KPI, and hourly level detail for the same KPI and another, when one data source contains information at a daily level and the other at hourly level, it becomes difficult to represent such information on a single dashboard. Necessary work-arounds must be employed to obtain the desired result. Generally, as rule of thumb, it is better to have the data at the lowest aggregation level so that it can be summarized in a dashboard and full drill down capabilities can be provided to the users for research and to derive actionable insights.

- **Establish distribution channel**: Dashboards are valuable if they provide actionable information, but also if they can be accessed by the user community when needed with minimum effort. To that end, determining the best distribution channel for the visualizations is critical. Since the visualizations are of immense benefit to various groups within the organization, at FAA, SharePoint – Knowledge Services Network (KSN) was deemed as the best distribution channel for this information. KSN allows the dashboards to be easily accessible and help retain most features of the business intelligence platform.

Air Traffic Operations Metrics (ATOM) is a collection of dashboards that are developed with a goal to help Air Traffic personnel monitor operations and make decisions to efficiently operate the National Airspace System. In addition to dashboards mentioned in this paper, the FAA has also implemented dashboards that provide details on Aircraft Diversions, Delays, and Special Use dashboards based on specific airport/facility requirements.

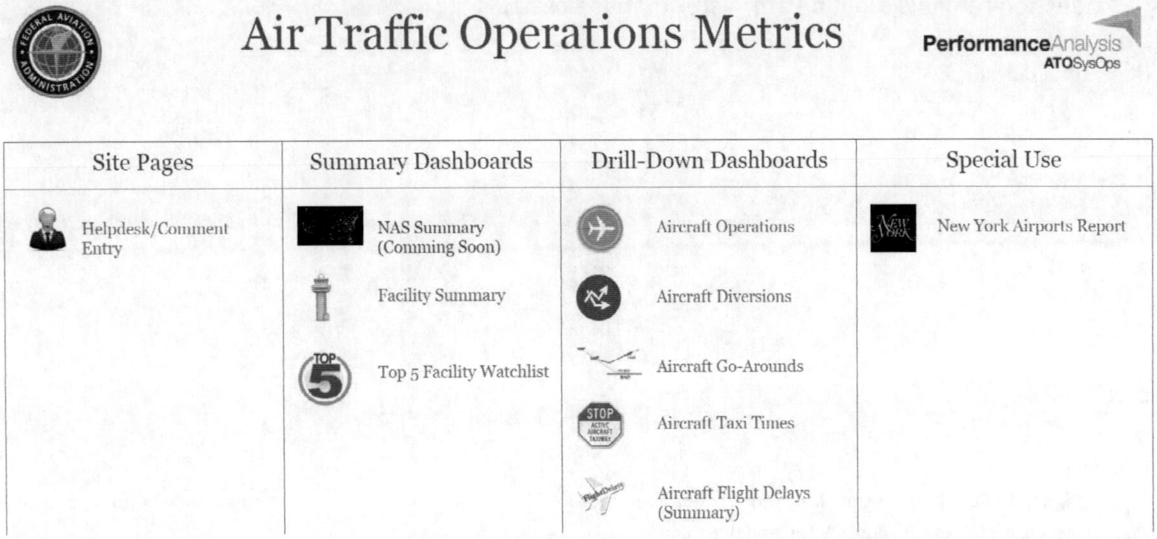

Figure 13: Air Traffic Operations Metrics dashboard homepage. A SharePoint site accessible by FAA staff.

Key Benefits

Key benefits observed while developing dashboards include:

- **Ability to blend data sources**: Blending is defined as ability to join data sources with different formats. Many business intelligence tools available today are data agnostic and connecting to a number of different data source formats. This has been immensely beneficial to dashboard development as up to 6 data sources have been used in some of the dashboards. Additionally, data in different formats can be blended, for example, blending Excel, with Oracle, with CSV files, thereby allowing dashboards to have a full drill down capability across multiple dimensions.

- **Ability to model within a tool**: Another benefit of business intelligence platforms is the ability to create parameters and calculated fields with complex calculations within the dashboard. Typically, when an analyst builds a model for a specific purpose, it is utilized for a specific period and retired. At the FAA, a dashboard was created for a specific airport to monitor KPIs and impact of a runway closure due to construction. A model was implemented within the dashboard that determines optimal runway configuration, ADR, and AAR given the wind angle and speed. Calculations were implemented to determine the head wind, tail wind, and cross wind component, which then was utilized in determine the runway configuration that can be used. This dashboard can be extended to other airports by making fairly minor changes to the model.

- **Time to market**: Historically a request to build a dashboard or a report would take weeks and may be even months depending on data availability and complexity of the calculations and visualizations.

This time has been cut down to days due to advances in business intelligence tools. As long as the data is available and is visualization ready, a dashboard can be constructed within a few of days. Below graphic shows trend for turnaround time in relation to visualization evolution. However, the question remains can it be any faster than it already is?

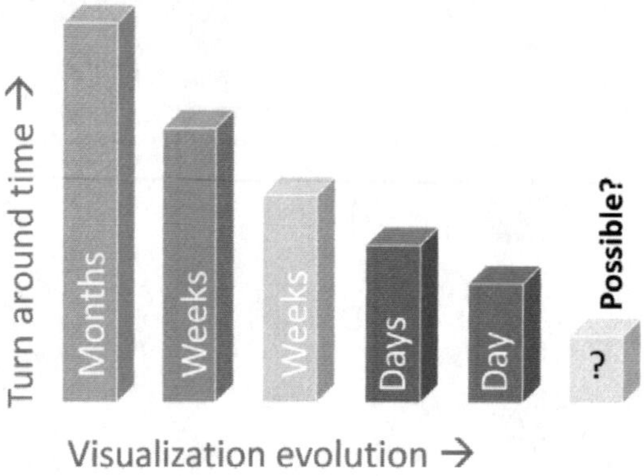

Figure 14: Dashboard Development Trend. Time required to develop dashboard reduced as visualizations evolved driven by advancements in business intelligence tools.

Actionable Business Intelligence – Untapped Space

Another advantage of business intelligence is their ability to integrate with statistical programs such as R. Tableau and other business intelligence products can embed an entire R script into the dashboard. This allows the script to be repeatable and can run in perpetuity as long as the data and necessary elements are available in a required format.

With R integration, the model and visualization becomes self-learning, as the model is able constantly take in new information, run analysis, and display results in an easy, understandable, and interpretable format.

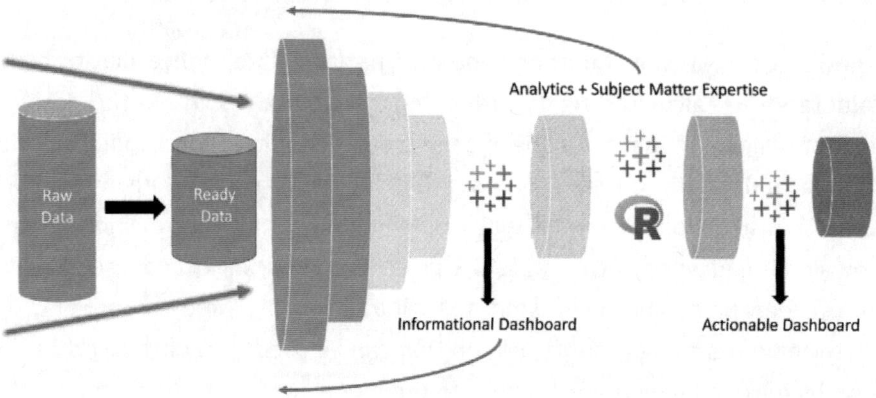

Figure 15: Informational vs. Actionable dashboard. Illustrates transition of an informational dashboard to an actionable dashboard.

The above figure illustrates the difference between informational and actionable dashboards. Also depicted in the above figure is the integration of R to develop actionable dashboard. Successful modeling of data with R not only requires cured data, but also required subject matter expertise in order to understand the data and ensure its relevance to the problem at hand. Large amount of data may be collected and ingested in the model. However, not all of it is significant and only significant data when further analyzed and visualized, results in an actionable dashboard that can be used to make operational decisions.

Aviation Application - Predicting Taxi-Out Times

A potential application of this concept could be to Taxi Times – specifically Taxi Out times. Taxi times are critical for operational performance, as they drive the capacity calculations for a facility. Determining how long it takes for an aircraft to taxi to the runway and wheels off determines the departure capacity of a facility. This is further complicated by inclusion of other factors including gates, terminal, equipment, carriers, and other operational factors.

Leveraging integration of statistical programs such as R with business intelligence and visualization applications such as Tableau, such model can be configured and implemented for each facility. Data can be pulled into and analyzed within the tool and the results can be visualized for easy interpretation by the user. Additionally because this model is embedded within the business intelligence and visualization tool having a constant input of data, this model becomes self-learning. As additional data is ingested by the model, the results are refined, thereby improving the accuracy of predictions of taxi times. Furthermore, any variables that are not significant can also be eliminated from the analysis. Significance of variables can be visualized in an easy interpretable format than in a statistical result table for easy understanding.

Lastly, the model can also be enabled to tag 'like' days depending on the 'like' criteria, so that the users can compare two days and analyze the differences that can be discussed and addressed. This will also enable users to adopt best practices across 'like' facilities for 'like' days in the NAS.

Ability for a facility to perform surface planning, delay planning, taking into account the schedule and slot control could help drive the decisions to efficiently operate the facility and in turn the National Airspace System (NAS).

Conclusion

The aviation industry collects massive amounts of data that is gathered using various methods and housed in disparate data stores. The industry relies on analytical models and systems that are built to consume data from these various data stores and output information to gauge operational efficiency. To date, access to such data and its integration with other data was difficult if not impossible, which in turn limited the types of analysis that could be performed.

However given the current era of big data coupled with advancements in business intelligence tools, we are now able to connect to various types of data sources via a single tool and blend them together to generate visualizations that help derive true actionable insights.

The FAA has adopted Tableau, a business intelligence platform that allows development and publishing of the interactive digital dashboards to be used across the agency. Similarly other agencies can review available tools and platform, and implement one that is best fit for them. Such tools allow creation of visualizations similar to the ones mentioned in this paper, and beyond. It enables users to perform secondary analyses and be more inquisitive about the data.

It is important to know that such tools are not just for generating reports. The true power of these tools, which allow creation of multi-dimensional dashboards, can be harnessed when it is used for research and analysis by generating interactive, impactful dashboards with full drill down capability and interactivity. Also, screen real-estate is limited, and therefore a careful attention must be paid to the amount of information included in the dashboards.

Some of the best dashboards take users through the funnel beginning at a summary level using specific simple visualization, through a set of visualizations increasing in complexity to more granular level. If a drill down is desired on the same screen, then left to right and top to bottom approach is recommended, where the summary visualizations are placed at the top left and the detailed visualizations at bottom right.

There should be some flexibility in data storage and retrieval strategy. Existing strategy may not be optimal for the dashboard exercise. At FAA, new schemas were setup and views were constructed to prepare data for visualization. Some data was pre-blended / joined before it was imported for visualizing. This helped improve the dashboard performance given that there were millions of records included in visualization. Additionally, data preparation steps such as harmonizing multiple sources, indexing, normalizing, extracting, transforming and loading are required. Once the data are prepared, it is considered visualization ready.

Leveraging business intelligence tools to connect to disparate data sources and developing multi-dimensional visualization to represent data can be immensely helpful in operational monitoring within aviation. Visualizations can be extremely powerful and convey a story in a clear and instructive manner. Users are more quickly engaged by seeing trends and outliers. By seeing the cause and effect of multiple factors, managers are able to more clearly forecast the effect of policies and drive decisions and operating practices that make the system more efficient and responsive.

Future Applications to Aviation

The business intelligence space is continually evolving specifically in developing self-service analysis and intelligence platforms for mainstream business users. These tools enable mainstream business users to conduct basis data analysis and generate data visualizations similar to the one illustrated in this paper.

Companies such as Tableau, which is a platform chosen by the FAA for business intelligence, have provided business users with the power to connect to, analyze, and visualize data on their own. Once the dashboard is developed it can be made accessible to various groups within the organization. The management can also determine the amount of data that needs to be made available for visualization. As long as the data is available, users can go back in time, perform comparative analysis and can derived actionable insights from the dashboards.

Some of the upcoming advancements in business intelligence platforms that can be applied to aviation are as follows.

Natural Language Processing:

Natural Language Processing (NLP) is a natural language interface that helps users interacts with data and visualizations. Users can ask questions in natural language and the result is a visualization responding to those questions. Example questions include: "How do the NAS operations in 2017 compare to 2016 YTD?" The expected result would be a visualization that shows NAS operations for January 2016 to April 2016 compared to NAS operations for January 2017 to April 2017.

A follow on question can be asked, for example, "How many Air Carrier operations were there between the specific time period?". In this case the expected response would be a revision of the visualization showing only Air Carrier operations for the above time frame.

Currently this technology / approach is available through specialized tools, which are programmed for a specific use case and have a predictive / machine learning component attached to it. But in the near future this will be in mainstream business intelligence platform. Users will be able to query the data, ask follow on questions, and the application will respond with visualizations that are pre-developed but displayed on the fly.

Integrating Media in Dashboards

Integrating media such as video, pictures, and simulations in dashboards is existing technology. However being able the queue media to a particular point upon selection of a specific measure would be an innovative way to help drill down into the details. This feature is on the roadmap for implementation in ATOM dashboards, especially the ones to be used for FAA executive briefings.

Acknowledgements

The author thanks employees of Federal Aviation Administration that provided input and permission to prepare and submit this paper. Additionally the author thanks the management and consultants of IT WORKS Inc. Special thanks to Dawn Amore, President, IT WORKS Inc., and Marlene Mandel, Program manager, IT WORKS Inc., for reviewing contents and providing feedback on the paper.

About the Author

Ankush Karnik is a Project Manager with IT WORKS Inc. Ankush has over 14 years of diversified experience in financial services and transportation sectors, performing functions ranging from digital marketing, social media listening analytics, consumer credit risk analytics, operations improvements, and managing capital and technology projects. At IT WORKS, Ankush is responsible for leading analytics and data visualization teams and engagements. Ankush continues to work with clients including the FAA to establish strategies to leverage analytics and visualizations to help derive both informative and actionable business insights. Also, currently Ankush is an adjunct faculty at the New Jersey Institute of Technology, teaching Engineering Economics and Project Management for engineering majors.

Introducing Data Model Standards for Data Comm Information Exchange over SWIM

Charles Chen
Antonio Correas

Skymantics, LLC

skymantics

Introducing Data Model Standards for Data Comm Information Exchange over SWIM

ABSTRACT

ATC services use datalink communication systems such as FANS/ACARS, FAA DCIS, and EUROCONTROL DLS/Link2000+ using dedicated Communication Service Provider (CSP) networks such as Rockwell Collins (ARINC) and SITA. A need has been identified in the broader aviation community for sharing up-to-date status information of ATC instructions and clearances with stakeholders. In order to efficiently disseminate this information, SWIM can be utilized to publish an XML-based data format and leverage existing publish/subscribe schemes. SWIM architecture can support the dissemination of the data and handle governance of the information to authorized users. However, a standard data model for this information is not yet specified. Introducing new "XM", called Datalink Messaging Exchange Model (DMXM), for datalink information sharing can ensure alignment with current international SWIM development efforts at the ICAO level.

DMXM can enable the use of a common reference data model for datalink information globally through SWIM by integrating it with the ICAO ATM Information Reference Model (AIRM). Aviation industry data exchange standards (e.g. ACRIS) can also benefit from the use of this information. Targeted users of applications for DMXM include airline dispatchers, pilots, airport operation control centers, and fixed-based operators such as ground handlers.

DMXM can provide the aviation community with a standard data model enabling ATC consumer applications through the use of interoperable semantics, thus achieving a higher productivity rate at a lower cost and a shorter time to deployment. Applications may support common situational awareness for controlled flights and enable collaborative decision making among stakeholders.

ATC ADVISORY SERVICES FOR DATALINK PROVIDE A MEANS FOR ADDITIONAL EFFICIENCY

ATM communications systems are currently segregated in two domains: safety and non-safety. The safety domain involves the exchange of command and control information and any data which supports safety of life and regularity of flight, whereas the non-safety domain involves the exchange of advisory information which supports operational efficiency but does not involve aircraft control. This segregation is applied to the ATSU, certain AOC service providers, and to the aircraft systems by means of physical separation of the Aircraft Control Domain and Aircraft Information System Domain.

Data link services are an integral part of safety-driven aviation implementations which use proprietary network services provided by communications service providers. The first implementations were ACARS-based FANS while, more recently, ATN is being used as the ICAO compliant standard to support ATC services. These data link services implement point-to-point dialogues over dedicated networks and interfaces managed by communication service providers using application-specific packet formats and protocols. On the other hand, SWIM is intended to provide advisory information (i.e. excluding any command or control exchange) and is based on service oriented architectures, standard messaging exchange patterns, and IP-based public/private networks.

Advisory information from SWIM can be greatly enhanced by correlating safety operations information. If data elements such as flight clearances, trajectory negotiations, avionics-sourced position reports, and in-flight weather reports are published to SWIM, then advisory information can be enriched with this data. Data link information distributed via net-centric information sharing using SWIM interfaces can improve collaborative decision making, situation awareness, and operations planning by ATM stakeholders. Users of the ground ATM network such as airports and AOC/FOC centers are the first to benefit from the continuous availability of ATC communications in order to support their own operations planning and execution.

ATM DATA STANDARDS LOWER COSTS THROUGH RE-USE

The standardization and adoption of new ATM data models is improving the way in which producers and consumers are sharing ATM information. Using XML, the producers of data are able to exchange information using SWIM which is rapidly decreasing the amount of time required to connect and acquire relevant data. Currently, SWIM is categorized as an efficiency-critical system, in contrast to safety-critical systems such as ERAM. This means SWIM systems and data can be used as a means of improving system efficiency of the ATM network and additionally for advisory information. While Data Comm information is considered safety-critical, the information produced by Data Comm Network Services can be published to SWIM for external consumption by airline operations such as dispatch and flight planning. However, publishing data in legacy formats causes challenges in consuming the data such as use of hexadecimal encoding and legacy nomenclature for point-to-point transport.

ATM data formats are defined in order to standardize the way in which ATM data producers and consumers exchange information. Adopting standards provided by OMG, ISO, and OGC can reduce the costs of implementation for SWIM technical services such as data validation and improve workflows, ensure data quality, and increase safety. The ICAO Information Management Panel (IMP) has introduced the concept of an ATM Information Reference Model (AIRM) for data standards over SWIM. However, the scope of Data Link information is unclear and does not currently fit in the AIRM. Therefore, defining a data model standard for this type of information in line with the others (AIXM, WXXM, and FIXM) should be considered if it is published to SWIM.

The proposed data model, called the Datalink Messaging Exchange Model (DMXM), is intended to provide information traditionally contained in legacy data link formats such as ACARS (ARINC 620 & ARINC 622) and FANS CPDLC (DO258A). DMXM needs also be designed in view of future support of ATN applications. Information published using DMXM benefits the airspace user community such as dispatch centers while providing additional information to enhance the SWIM data feeds for post-operations such as metrics calculations for improved operational efficiency.

DATALINK MESSAGING TYPOLOGY IS FRAGMENTED

Current datalink communications are driven by standards that mandate specific application functions over specific packet formats and message interfaces in order to be certified for operation in safety-domain networks. The phraseology for permitted ATS services is described in the PANS ATM Manual (ICAO, 2007), and communication and networking protocols must be compliant with ARINC, RTCA, and EUROCAE standards. These define different communication stacks per application type, which results in the fragmentation of datalink standards. The following existing standards are currently in use:

Aircraft Communications Addressing and Reporting System (ACARS) is a character-based messaging protocol for air-to-ground services using specific Communication Service Providers (CSP). ACARS core messaging is defined in ARINC 620 (ARINC, 2014), together with the conversion rules between air/ground, CSP, and ground segments.

ARINC 623 (ARINC, 2005) specifies character-oriented ATS messages (also called pre-FANS) that support D-ATIS, Departure Clearance (DCL) and Oceanic Clearance (OCL) applications. Although quickly superseded by FANS, character-oriented applications have been widely implemented by making use of the ACARS network for AOC services.

Future Air Navigation System (FANS) 1/A+ services standardized by ARINC 622-4 (ARINC, 2001) also use ARINC 620 core messaging, but support bit-oriented ATS applications, introducing error detection and addressing the interoperability among regional implementations as compared to ACARS.

RTCA DO-258A (RTCA/EUROCAE, 2005) specifies three defined ATS FANS applications: ATS Facilities Notification (AFN), controller to Pilot Data Link Communications (CPDLC), and Automatic Dependent Surveillance (ADS). FANS covers oceanic routes globally as the communication performance adequate for Oceanic and Remote use. In addition, the FAA Data Comm program has chosen FANS 1/A+ for use in domestic operations, including the introduction of Departure Clearance (DCL) in 2015, and plans to extend CPDLC services for enroute operations in 2019.

Aeronautical Telecommunication Network (ATN), also called FANS 2/B+, is a communication architecture developed by ICAO to provide a global air-ground and ground-ground data link applications and a communications standard for ATS. ICAO Doc 9880 (ICAO, 2014) defines the ATN air-ground applications as CPDLC, CM, ADS-C and FIS.

The main difference in ATN compared to FANS is the existence of a Dialogue Service (DS), which allows end points to communicate in a standard way, agnostic to the application, and supports either connectionless or connection-oriented communication services underneath.

The first implementation stage of ATN is called ATN Baseline 1 (B1) services (RTCA/EUROCAE, 2007) (RTCA/EUROCAE, 2012) which include CM and CPDLC, currently deployed in Europe following the Data Link Services Implementing Rule (DLS-IR) over VDL Mode 2. Future ATN Baseline 2 (B2) [9, 10, 11], includes additional applications supporting the all of the ICAO ATN applications that enable 4D trajectory based operations.

This fragmentation of datalink network and application specifications makes it challenging for the aviation community to access to ATC data communications in a re-usable, cost-effective way. By making it available via SWIM in a standardized format, Data Comm operational information can be decoupled from the message phraseology and typology and simplify processes using a common data model.

DATALINK MESSAGING EXCHANGE MODEL SUPPORTS INTEROPERABLE EXCHANGE FOR FUTURE CDM AND TBO

As future data link standards evolve, the information becomes more accessible through upgraded ATC systems, especially through the use of SWIM networks to disseminate and exchange information. For example, the FAA DCNS currently aggregates all the data link messages that flow between ATC and aircraft through the Tower Data Link Services (TDLS). This information is decoded and forwarded to the corresponding Rockwell Collins or SITA CSP for processing. Meanwhile, a copy of the message can also be copied and decoded as an XML data message and published on SWIM to be consumed by authorized airspace users. Dispatch offices and Airline Operations Centers (AOC) can use this data to monitor ATC to Pilot communications as well as compute metrics based on ATC-to-Pilot interchange. As this information evolves, the combined information of flight clearances and trajectory information with flight plans, aeronautical charts and airspace information (NOTAMs and SAAs), and meteorological data (PIREPS, Altimeter settings, common winds) with SWIM weather products can help airspace users perform more efficiently through use of SWIM.

The TDLS messages are currently aggregated by SITA using legacy formats and major airlines are currently paying a fee for use of this bandwidth to consume this information. The first indication when presented with the concept of a SWIM feed was overwhelming support from the airline community and dispatch user at large. Many of the major airlines already have connections into FAA SWIM, and providing a source of data link messages such as FANS and ACARS information through already existing channels would be an efficient means of leveraging existing infrastructure.

One case study is the use of DO-258A as described for FANS messaging. For example, a complex data element described in DO-258A is the ***routeclearance*** element. This element describes a sequence of data structures defined in the AEEC 702 Characteristic needed to provide a route clearance, and provides additional data elements in a complex nesting structure. Through the use of an XML schema, these data structures can be well defined and do not require human readability to process. The data syntax can be enforced and validated prior to transmission and upon receipt. Using modern software methods, XML is easily converted into data objects (e.g. Java JAXB) and loaded into a database for immediate processing or correlation with other SWIM data elements.

In Figure 1, the Common section of the message contains information on the SWIM message as produced by the SWIM publication services such as related metadata such as operator, timestamp, and messageType. The Datalink Transport section will contain the extracted datalink transport information such as ACARS, FANS 1/A, and ATN. For example, these elements may include A620 and/or A622 fields from the FANS message. The message payload will contain the command and control message from ARINC 623, DO-258A, ICAO 9880 formatted sources, for example, an uplink message or downlink message in DO-258A format transformed to XML.

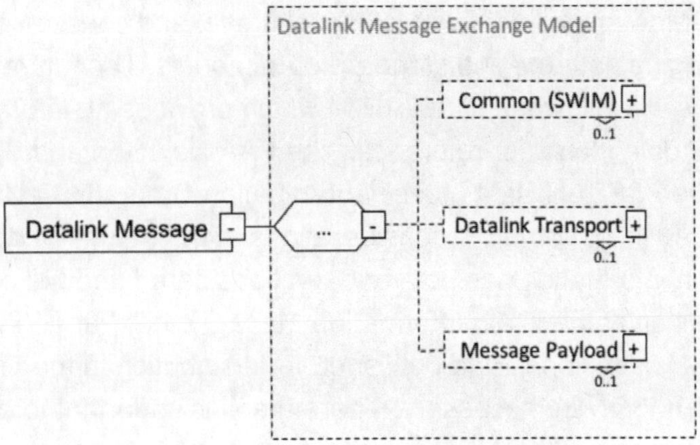

Figure 1. DMXM Message Structure

CURRENT SWIM ARCHITECTURE CAN SUPPORT DMXM

Figure 2 depicts the overall architecture of the datalink message concept in the US National Airspace system (NAS). ATS datalink messages flow between ATSU and aircraft cockpit systems (e.g. FMS) through an ATC gateway operated by the CSP. A copy of the message is created, translated into a DMXM-compliant XML file and published to NEMS using the Java Message Service (JMS). Then this message can be consumed via SWIM by subscribers to the service, whether they are located in a ground network (e.g. airline AOC) or in the aircraft (e.g. EFB). In the latter case, an IP air-ground datalink subnetwork with bandwidth, QoS and security capabilities is required, as described later in this section.

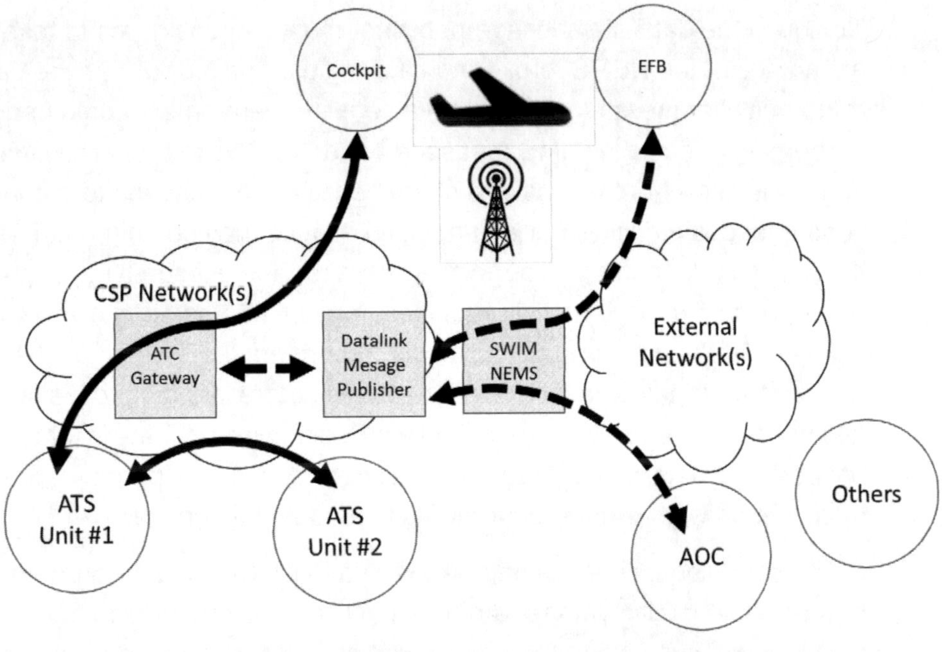

Figure 2. Architecture showing ATC service communication (straight lines) and ATC advisory publication to non-ATC users (dotted lines)

The DMXM must begin with a logical model and consider the current SWIM messaging patterns and standard data models. The future of these standard data models will follow the ICAO AIRM which provides an information model and a logical data model for reference. The FAA NextGen and SESAR have formed the NextGen-SESAR Data Model Coordination Group (NSDMCG) to develop the ICAO AIRM governance considerations [14]. The use of data link and safety-domain information has not yet been considered and will introduce some level of discussion at the ICAO Information Management Panel (IMP). Assuming that DMXM is developed and published for distribution in the future, alignment with the AIRM may be necessary. Therefore, it is important to consider design specifications of the AIRM conceptual model when formulating DMXM constraints and representations. One example is the use of EPSG:4362 Coordinate Reference System (CRS) for representation of position and geospatial references which is used in AIXM, FIXM, and WXXM models.

SWIM provides security control functions such as authorization, authentication, and encryption. For example, the FAA NAS Enterprise Messaging Service (NEMS) acts as a broker between NAS data publishers (such as the publisher of datalink messages) and external consumers of SWIM. Every SWIM consumer must pass a series of stages to "on-ramp" to various NEMS testing environments as a qualification to be approved to connect to the operational environment. First, a user must connect using a secure virtual private network (VPN) connection to the FAA Telecommunications Infrastructure (FTI). Once connected, a user must authenticate using a Java Messaging Service (JMS) client applications. Finally, the user can receive information on their designated JMS queue/topic only.

Design of the datalink message publisher must be considered to take full advantage of the NEMS' brokering capabilities. The NEMS has the capability to route and filter messages based on a series of parameters contained in the JMS properties (often called JMS message headers). These properties may be used along with a NEMS taxonomy to characterize each message to a corresponding consumer desired filter. For example, messages may contain headers such as the airline code, departure airport, arrival airport, message type, flight ID, and so on. A consumer, such as Delta Airlines, may be interested in all messages for flights arriving and departing in Atlanta airport, but the messages they receive will also be filtered to contain only the messages corresponding to Delta Airline's airline code. A Java Messaging Service Description Document (JMSDD) [15] provides the details of the message headers and taxonomy as it applies to a JMS publisher as well as the requirements related to the service.

A request/response messaging pattern may also be used for on-demand requests for specific data regarding an active flight or airspace boundary. In order to implement such a service, use of web services is required. A web service requirements document (WSRD) [16] describes the characteristics of a web services designed for the business rules and operations and the requirements of the services.

In addition, attention should be paid to the infrastructure supporting the message exchange services described herein. IP-based ground aviation networks (e.g. FTI in the U.S., PENS in Europe) are the likely candidates. The design of DMXM is expected to drive requirements on the underlying communication architecture to transport datalink messages to the authorized users. The following capabilities should be supported by the network components and hosts to guarantee an acceptable level of service:

- High capacity and QoS,
- High service and network availability,
- Robust security features: authentication, confidentiality and integrity,
- and Prioritization

AIRSPACE USERS BENEFIT FROM EFFICIENT RE-USE OF TOOLS AND APPLICATIONS

The use of a standard data model for data link information provides a single mode for transporting all data link information in a standard feed. Airspace users may use the same software tools to consume FANS data as they do future ATN-B2 data as long as it retains the same common data elements. Vendor lock-in is reduced and airspace users (and others, such as airport stakeholders as described) can invest in augmenting current tools without worrying about technology refresh due to new data Extract Transform and Load (ETL) requirements from a new data link technology.

In order to integrate seamlessly IT solutions from airport stakeholders and airspace users as datalink message service users, the design of DMXM should consider interoperability with the ACI Airport Community Recommended Information Services (ACRIS) and the IATA Aviation Information Data Exchange (AIDX) system. To do this, services incorporating DMXM can define component identifiers to be instantiated as ACRIS and AIDX-compliant attributes. This will create uniquely identifiable components for airport stakeholders and airspace users to facilitate resource reuse in specific business application development.

By using SWIM to transport data to ground networks, dispatch systems can capture controller-to-pilot communications in real-time and intervene as necessary to assist pilots. This functionality can be further extended in the future with the transmission of relevant information directly to the pilots by air-ground datalink. Electronic Flight Bags (EFB) are already being introduced in the cockpit to provide pilots with support advisory information, and built-in app clients are able to consume SWIM information. Visualization tools that consume and process appropriately filtered ATC information of flights in the vicinity enable a completely new paradigm of autonomous flight separation and trajectory conflict detection systems. This can also improve collaborative decision making processes and eventually support trajectory based operations once services are hardened per safety requirements.

Post flight analysis on data link information can provide dispatch and flight planning offices at the AOCs to improve flight operations and efficiency. Opportunities exist here to apply big data analytics algorithms on to assess cost impact of ATM over operations, and human factors impact to ATC compliance.

FUTURE WORK

Further work must be conducted to translate legacy formats into DMXM and collaboration with industry is of utmost importance. Current development of SWIM services to support data link messages is already underway, and the benefits will be captured by industry and evaluated for improvements.

As new data link standards emerge, a paradigm shift will occur to support trajectory based operations. Uplink and downlink of trajectories for negotiation, uplinking of weather and aeronautical updates, and downlinking of Extended Projected Profiles (EPP) will continue to change the way ATM operates, and sharing of this information with authorized airspace users and dispatch will close the loop on Collaborative Decision Making (CDM). Trajectory Negotiation, Advanced Interval Management, Dynamic RNP, ADS-C Extended Projected Profiles, ATC Winds are some examples of TBO concepts that can benefit from the use of a DMXM data feed. If designed properly, DMXM can support forward compatibility for future datalink services to support 4D trajectory negotiation and execution.

Works Cited

ARINC. (2001). SPECIFICATION 622-4. *ATS Data Link Applications over ACARS Air-Ground Network.*

ARINC. (2005). SPECIFICATION 623-3. *Character-Oriented Air Traffic Service (ATS) Applications.*

ARINC. (2014). SPECIFICATION 620-8. *Datalink Ground System Standard and Interface Specification (DGSS/IS).*

ICAO. (2007). Doc 4444 ATM/501 Procedures for Air Navigation Services. *Air Traffic Management, 15th.*

ICAO. (2014). Doc 9880 AN/466 Manualon Detailed Technical Specifications for Aeronautical Telecomminication Network (ATN) using ISO/OSI Standards and Protocols. *Draft, 2nd.*

RTCA/EUROCAE. (2005). DO-258A/ED-100A. *Interoperability Requirements for ATS Applications Using ARINC 622 Data Communications (FANS 1/A Interop Standard).*

RTCA/EUROCAE. (2007). DO-280-B/ED-110B. *Interperability Requirements Standard for Aeronautical Telecommunication Network Baseline 2 (ATN B1) Interop Standard.*

RTCA/EUROCAE. (2012). DO-305A/ED-154A. *Future Air Navigation System 1/A - Aeronautical Telecommunication Network Interoperability Standard.*

skymantics

Charles Chen

Charles is a co-founder and CEO of Skymantics, LLC, an aviation consulting firm. He is a subject matter expert in System Wide Information Management (SWIM) research and development with a focus on the Flight Information Exchange Model (FIXM) and Flight Object concepts. He provides SWIM/SOA architecture support for multiple industry testbeds including the OGC Testbed, the FAA Florida NextGen Test Bed (FTB), and the NASA SMART NAS testbed.

Charles has a Bachelor's degree in Electrical and Computer engineering from Auburn University and a Master's degree in Systems Engineering from Florida Institute of Technology.

Antonio Correas

Antonio is a co-founder and President of Skymantics, LLC. He is a specialist in FANS/ATN data link products for air traffic services and airline operation communications, air-ground communication subnetworks (AeroMACS, VDLm2, SATCOM, UAS CNPC) and IP aviation safety services. He has experience in supporting commercial and government customers worldwide in the development of standards and market adoption of new wireless technologies to enable new air traffic management concepts of operation.

Antonio is a graduate of Embry-Riddle Aeronautical University with an MBA in Aviation. He holds a B.S. in Telecommunications Engineering from Universidad de Zaragoza, Spain and a specialization in the Master of Wireless Networks from Kungliga Tekniska Högskolan, Sweden.

Opportunities and Challenges for Use of UAS at Airports

By Suzette Matthews Esq., Frank L. Frisbie P.E., and James H. Cistone, Ph.D.

A recognized advantage of Unmanned Vehicle Systems (UAS) is that many will be able to operate outside our limited and overburdened system of airports. There are, however, at least three categories of opportunities for use of UAS at airports.

Opportunities

First, airports and their subsidiaries and service providers can themselves use UAS at the airport and in its environs to improve the quality, efficiency, and economics of their own operations. Among those jobs that UAS can perform around the airport are surveillance for safety including fire and rescue, security and law enforcement, and facilitation of airport operations such as ground vehicle traffic management, collision avoidance, wildlife management, environmental monitoring, surveying (i.e. Obstacle and Terrain evaluation), and even small package delivery within the airport confines.

Second, there will be some UAS operators who will become customers of the airport. Although most small UAS operations probably will take place outside the airport environs, airports will eventually host some significant UAS operations. For example, it can be envisioned that small package delivery companies might operate large aircraft into airports, and then use UAS to pick-up and dispatch small packages from the airport directly to customers in the local environs. Direct dispatch from the airport would potentially save time and expenses associated with local warehouse, sorting, and road delivery infrastructure. And over time, it may well be that military and many large aircraft cargo carriers already residing at airports will convert to optionally-piloted and unmanned aircraft. There are numerous joint use military and civilian airports where, and from which the military may want to operate drones or expand current drone operations.

Third, it is likely that UAS or combined UAS and manned aircraft airports will be newly built exclusively to serve UAS customers. For example, a distributor or group of distributors might build distributions centers in an industrial park locations, around and from which they could deliver goods by UAS.

Challenges

There remain significant challenges however for use of UAS at airports, many of which are not, or are not adequately being addressed.

Some of these challenges are technical. Large UAS such as Global Hawk, properly fitted with avionics (e.g. voice communications, ADS-B, GPS) may operate and be handled by ATC indistinguishably from manned aircraft. Small UAS however may well operate largely outside the traditional ATC system, raising the following safety and operational issues:

- How will airborne surveillance of small UAS be assured, both for ATC and other operators when most will not be position reporting to ATC? Will all UAS in and around airports be required to have transponders? Will other operators be required to have some sort of primary surveillance and collision avoidance capability to avoid such small targets or will avoidance always be the responsibility of the UAS? Would large aircraft operators—and their passengers—accept the idea of UAS having sole responsibility for avoidance? What surveillance, if any, will be required for surface operations?
- Will small UAS operating in and around airports, including surface areas be subject to some sort of positive control for anti-collision or airport efficiency, and if so, who will perform this function and

how? What infrastructure will be required, such as information and communications systems and air traffic management tools? If there is some form of positive control, what will be the control procedures and prioritization paradigms in situations of high or excess demand, or when airspace, runways, landing areas or parking is limited or capacity is constrained. To what extent would segregation of airspace and surface operations suffice?

- What is the minimum performance capability required of UAS in particular terminal environments for specified operations, and can these standards vary depending on where and how the UAS will be used? Should these characteristics, and/or physical characteristics, be circumscribed such that UAS can receive uniform treatment?
- Will radio spectrum be sufficient and if not, how will this be addressed?
- What will be the certification and safety standards for UAS operating in and around airports, and who will perform certification and safety assurance functions? It is not a forgone conclusion that FAA can or will assume this role with respect to small UAS operating at low altitudes.
- What failure contingencies will be established to cover all potential UAS failures (e.g., lost ATC communications, navigation, or surveillance, inability to continue flight, or declared emergency)?
- How can risk to the public of UAS operations at airports be effectively analyzed and assessed?
- Will UAS operations pose environmental concerns, including noise?
- How can the airports and their customers assess the operational and financial impact of UAS operations on their current business, including liability?

Other issues will have policy, legal, and public acceptance implications, such as:

- What level of risk to the non-participating public from UAS operations is reasonable and acceptable, and by whom and how will this risk standard be set and enforced?
- How would the public be protected from residual risk? Should third-party insurance be required of operators, and if so how might appropriate coverage amounts and liability limits be established? By whom and how should compliance with insurance requirements be enforced?
- Do airport customers, suppliers, and employees, or neighbors of airports have any expectation of privacy that might be infringed by UAS operations, and if so what are the boundaries of these privacy rights, and how might invasions of privacy be avoided, mitigated, or compensated? To what extent could public outreach improve the chances of public acceptance?

The most significant immediate challenge however for airports and their customers will be to secure FAA safety approval for the most useful applications of UAS. At this point, the boundaries of authorized UAS use are very narrow. FAA has attempted to accommodate the most pressing UAS operator demands through exemption processes (so called "Section 333 Exemptions"), and now through new FAR Part 107 (14 CFR Part 107). But both the exemption process and Part 107 prohibit commercial UAS operations within 5 miles of an airport, require the UAS to fly under 400 feet, only within the human UAS pilot's un-aided line of sight (LOS), and only over people participating in the operation. Neither avenue of approval allows the operator to fly UAS near an airport, over the public, beyond the human capability to see the vehicle (beyond line of sight or BLOS), or using "autonomous" systems such as pre-programmed flight routings. Absent the 5-miles restriction, such limiting operating parameters might suffice for some on-airport uses of UAS (e.g. surveying or wildlife monitoring), but ultimately more sophisticated capabilities will be integral to most practical business and commercial applications.

FAA has invited prospective UAS operators to apply for "waivers" of the requirements of Part 107, or to use its general exemption processes to apply for certifications/approvals of UAS operations beyond those already permitted. To earn a waiver or exemption, the application would have to seek approval of an alternate method of compliance (AMOC), having demonstrated that the proposed UAS operation achieves an equivalent level of safety (ELS) with the analogous manned operation. FAA has provided little guidance on how to make such a showing, and an applicant attempting to do so faces the challenge of baselining and measuring the safety of current operations in the absence of good historical and current data. As of the date of this writing, FAA has granted very few waivers from the over-the-public prohibition, and none with wide ranging practical, much less commercial, application.

Addressing the Challenges

The authors believe that a comprehensive program leading to full UAS integration in the NAS is necessary to address all of these challenges, including those inherent in UAS operations at airports. The scope and elements of such a program were presented in a paper and presentation to AUVSI's 2015 annual meeting in Atlanta. The full paper "An Achievable Path to UAS Integration in the NAS" can be read and downloaded at http://www.safeaccess4uas.com/paper-uas-integration-nas.html. Although the paper describes the requirements for full integration of UAS into the NAS, *the same technical and policy elements would be necessary to safely and acceptably integrate UAS operations into the microcosm of an individual airport:*

- Real time position determination, reporting, and information sharing for all vehicles (not just UAS (where we all are)
- Intent information (where everyone is going)
- Some form of positive control over everyone (collision avoidance), even if only by exception, and an airspace management paradigm and mechanism to prioritize operations if necessary
- Adequate spectrum, which will be a combination of government allocation (and preservation of the aviation bands) and technical advances to divide and better use what spectrum there is
- A performance-based safety certification, approval, and assurance regime, including articulated standards (acceptable risk to the public) and measurable outcomes, whereby the performance of diverse UAS in various operating environments can be demonstrated, measured, and adjudged objectively
- Methodologies and tools for assessing residual risk to the public from UAS operations, and requirements for compensating the public for injury and damage from mishaps (e.g. insurance coverage or adequate financial wherewithal for self-insurance)
- Analysis and prevention and/or mitigation of any unacceptable invasions of privacy or civil rights inherent in specific UAS uses at the airport or to airport neighbors.

Needed most of all however, is active acknowledgement and acceptance by the entire aviation community of the fact that UAS will be full participants in the NAS, and for some entity—be it government or aviation community initiated and driven—to lead, drive, and manage the technical and policy activities toward that objective. At this point, FAA has not accepted this role and no other organization—public or private--has come forward to fill the gap.

Impacts of UAS Operations at Airports

It is possible that some airspace, surface, and terminal building areas at the airport could be sufficiently segregated and sanitized so as to allow some small UAS to operate without disrupting routine airport operations. But to the extent that UAS attempt to integrate into routine airport operations, they will impact virtually everyone there.

On the positive side, airports and their service contractors can realize operational improvements (e.g. speed and comprehensiveness) and economies by using UAS for what are now labor intensive activities such as observation and surveillance of airport operating areas and wildlife, and environmental or other potential hazards, and airport surface management. Within the terminal structures, UAS have the potential to improve security observation and surveillance, and might even perform rapid document and small package delivery. On the negative side, airport customers including passengers may object to the sense of being constantly observed (although observation cameras and recordings are pretty much ubiquitous these days, performing pretty much the same function.)

Some current aircraft operators, such as large cargo and small package delivery companies, may contemplate significant business opportunities in UAS operations at airports, including the possibility of UAS pickup and delivery of packages and small cargo beyond the airport. And there may be a rush on building new airport facilities, new UAS airports, or hybrid UAS/manned airports to serve these needs.

On the other hand, other aircraft operators at the airport—commercial and general aviation--may be negatively impacted. Most will not have on board primary surveillance equipment or the maneuverability needed to detect and avoid small UAS, which will be perceived as a safety hazard even if the UAS themselves are capable of detecting and avoiding the manned aircraft. And to the extent UAS are capable and desirous of using the airport under ATC management and control, they will add demand to airports already experiencing congestion and/or demand management.

All of these stakeholder groups are represented by trade organizations, such as AUVSI for UAS manufacturers and operators, A4A for airlines, NBAA for business aircraft operators, AOPA for general aviation pilots and companies, NATCA for air traffic controllers, AIA for aerospace and avionics manufactures, and AOCI and AAAE for airports and airport management, etc.

Addressing and Resolving Issues

As discussed above, airports and their customers may explore limited UAS operations at or around the airport under new FAR Part 107, namely those tasks the UAS can perform while being actively flown within direct line of unaided sight of the UAS pilot, and not flown over the general public. To the extent the airport or its customer can demonstrate an acceptable level of safety, some limited expansion from the Part 107 restrictions may be obtainable through the waiver and AMOC processes. If there are airport customers who believe they can operate larger, technically sophisticated aircraft under ATC control in a manner that is indistinguishable from manned aircraft, safety certification consideration under FAA's UAS Pathfinder program may be worth pursuing.

Airports may consider creating segregated, sanitized airspace and surface areas exclusively for UAS that do not conflict with routine operations of manned aircraft and ground operations. Safety risk analysis would have to be conducted to assure non-confliction with regular operations under all circumstances, including bad weather, emergency, security breaches, or other contingency situations.

Meanwhile, NASA is engaged in a public-private collaborative program to enable all nature of UAS operations in airspace below 500[1] feet—the UAS Traffic Management System (UTM). Even assuming the technical success of UTM however, under-500 feet is precisely the environment rife with privacy and public non-acceptance concerns. Geo-fencing of the airspace above non-consenting property owners has been suggested as a way of mitigating landholder resistance, but this would create a patchwork of authorized airspace that might prove operationally impractical. And the need for UAS to transit non-controlled airspace between UTMs, and for other marginally-equipped aircraft such as helicopters and general aviation to impinge on UTM airspace raise some very vexing detect and avoid issues.

[1] Title 14 CFR Part 107 currently limits altitude to 400 feet

WPG has recommended that rather than focus on incremental solutions, the most practical approach would be a comprehensive, aviation community-wide activity to integrate UAS fully in all aspects of the NAS.

Putting Potential Solutions into Practice

Taking into account the many uncertainties and challenges associated with UAS operations at an airport as outlined herein and in prior TRB work, the best approach might well be a limited experiment from which quantifiable costs and outcomes can be derived. And it will be even better if the experiment is directed toward demonstrating a viable (e.g. cost effective) business case. For example, an airport might explore with a package delivery firm already resident at the airport the concept of using UAS to dispatch small packages destined for the immediate locality, rather than by ground transport. Such an experiment would ferret out many of the issues that would be associated with more extensive UAS operations from an airport including unanticipated costs, the nature and extent of benefits, regulatory barriers (not only federal, but state, local, and airport), and potential liability and community acceptance issues.

Preparatory assessments and approvals for the experiment would include at the least:

- An assessment whether the planned operation can be performed within the confines of FAR Part 107, or whether waivers from those conditions (e.g. requiring LOS, and precluding operations over the public) will be required
- Whether the airport's physical plant, the operator's facilities, and ancillary services such as ATC and insurance coverage options are adequate and available
- Whether the UAS operator can perform the operation safely, both on and beyond the confines of the airport, including the last 100 feet to the delivery point, which is proving technically problematic
- Whether other customers of the airport especially other aircraft operators will object to the operation, and if so, whether their concerns can be addressed or mitigated
- Whether and to what extent the operation will be met by public non-acceptance, and if so, whether and how this discomfort can be addressed or mitigated.

Suzette Matthews, Principal, Washington Progress Group LLC, an aviation attorney. She represented major foreign and domestic airlines before the CAB, FAA, DOT, and Federal Courts. She has served as Executive VP and General Counsel, and Director of the Air Traffic Control Association, and Editor of ATCA's *Journal of Air Traffic Control;* was a Senior Subject Matter Expert to the FAA Joint Planning and Development Office (JPDO). She is the author of numerous published articles on aviation law, technology policy, unmanned aircraft issues, and public-private partnerships. She received the *Aviation Week and Space Technology* Laurel Award (2001), the Air Traffic Control Association Clifford Burton Award (2004) and Chairman's Citation of Merit (2012). She holds a B.A. With Distinction and Phi Beta Kappa from Cornell University, and a Juris Doctor degree from Cornell Law School. She is a Member of the Bars of Virginia and the District of Columbia.

Frank L Frisbie, Aviation consultant, former FAA and industry NAS Senior Executive. Frank has 55 years of experience in ATC/ATM spanning the full design, development, implementation, sustainment and replacement cycle of all NAS infrastructure elements, including a professional career in FAA culminating in the position of NAS Program Director and Acting Associate Administrator of Development and Logistics. Frank also served as a senior executive of Northrop Grumman. Frank holds a Bachelor's Degree in Electrical Engineering (BEE) from Manhattan College, NY and a Master's Degree in Business Administration from American University, DC. An Honorary Member of ATCA and recipient of the Glen Gilbert Award, Mr. Frisbie is a frequent contributor to the Journal of Air Traffic Control.

James Cistone, Ph.D., is a Senior Aviation Subject Matter Expert with Sullivan Aviation Services, LLC. With more than four decades of experience, Dr. Cistone is a proven engineering and business professional. He has supported software systems development and integration and the complete product/program life cycle including domestic and international business development and capture as well as engineering and program management, earned several advanced degrees including a doctorate in aviation, authored many published works, presented as a subject matter expert at industry conferences and conventions, and taught graduate level courses at Embry-Riddle Aeronautical University.

A White Paper Redefining Collaborative Air Traffic Management (CATM)

By: Jeff Burke Ph.D. (ABD)

Disclaimer: Even though the FAA employs the author, the information provided in this White Paper is strictly the perspective and opinion of the author and does not constitute a change in policy for the FAA.

NextGen defines Collaborative Air Traffic Management (CATM) as a philosophy that delivers information and flow decision-making data to flight planners and FAA traffic managers with the goal of improving the overall efficiency of the National Airspace System (NAS). Also, NextGen's CATM endeavors to provide greater flexibility to the flight planners and make the best use of available airspace and airport capacity. The overall philosophy driving the delivery of CATM services is to accommodate user preferences to the maximum extent possible. (NextGen, 2015).

As NextGen defines CATM, CATM does not obligate stakeholders to participate in the collaboration process, CATM only encourages the collaboration process. However, as technology advances so does the delivery of information, and so does the ability to collaboratively make decisions in a data-rich environment. As this white paper will describe, it is time to advance the philosophy of information sharing using online collaboration utilizing visual analytics. While promoting the philosophy of how decisions are made in a collaborative environment, I will propose a paradigm by allowing stakeholders the ability to define and express *their* NAS management criteria and performance goals, and communicate these criteria and goals to traffic managers. However, more importantly, the stakeholders would be able to utilize real-time data and information as they compare *their* NAS management criteria to historical data to get a feel for the results of *their* decisions. The information would allow the stakeholder to determine if it meets *their* goals and communicate their decisions to traffic managers. In today's environment, traffic managers define goals, then develop initiatives and implement the initiatives that attempt to meet those goals. That is archaic thinking, and it needs to change and here is why.

PART 1

The Need for a new Collaborative Air Traffic Management (CATM) philosophy.

PART 2

Existing Technologies that will be critical.

PART 3

Existing Philosophies that are utilized in today's traffic management environment.

PART 4

A New Paradigm based on system thinking instead of analytical thinking.

PART 5

We Have the Data! If the data is not available, then it does not matter.

PART 6

How Do We Do It? That is an intimidating question, or it?

PART 1 THE NEED

In today's data-rich environment, stakeholders have the need and the desire for accurate and timely information. They require this information to make informed business decisions that affect hundreds and thousands of their customers on a daily and even hourly basis. However, due to a variety of reasons, we see a decline in the ability to comprehend and understand the mass volume of information that is out there. That is one of the foundational pillars of the Collaborative Decision Making (CDM) process. The thought process is that if everyone has the same information that the group would be able to develop the solution to any given issue. However, we quickly determined that (1) there can be multiple solution sets for the same issue, and (2) that business models and agendas can undermine the collaborative process.

The need is to develop a communication conduit to deliver the information and allow the stakeholder to define *their* NAS management preferences and criteria, to enable them to witness the results of their future decisions, and to let them determine if it meets *their* needs.

As stated, in today's collaborative environment the goals are defined, and decisions developed that move towards or meet those goals. That **NEEDS** to change.

What is necessary is for the traffic manager to be able to insert the stakeholder's NAS management preferences or criteria (operational goals) into the center of the decision-making process, enabling traffic managers to prioritize which initiatives or ideas are the best way to meet the system's needs and goals.

As a point of discussion, this white paper will define an operational goal as focus points for NAS operations, the goal drive action towards an outcome.

The objective of the white paper is to identify a way to develop operational goal by utilizing the collaborative power of the NAS. The **need** is to define and develop operational goals and give them meaning. However, the goals must be measurable, but more importantly, it must be determined if is it the right goal. That is where Collaborative Air Traffic Management comes into play.

PART 2 EXISTING TECHNOLOGIES

Aviation System Performance Metrics (ASPM)

Aviation System Performance Metrics (ASPM) is a database maintained by the FAA that collects data each day for 77 airports. Some of the data that is collected by ASPM: OPSNET delays, departure cancellations, arrival cancellations, average taxi-out times, departure delays, the number of airborne holding events, airborne holding minutes, diversions, go-arounds, and completion rate.

Terminal Aerodrome Forecast (TAF)

The Terminal Aerodrome Forecast (TAF) is a weather-reporting format that forecasts weather at an airport every six hours. The forecast applies to a 24- or 30-hour period, and an area approximately five statute miles from the center of an airport.

Airport Decision and Support Tool (ADEST)

In today's traffic management environment, airport arrival rates, many times are only the opinion of the traffic managers, of how well the system will perform based on their experience and knowledge. The 7210.3 does offer a formula to assist in determining arrival rates (airspeed of the aircraft over the threshold of the runway divided by the number of miles between aircraft). However, there are many more factors used to determine the arrival rate of an airport; a great example is the runway construction at LAX in 2016 and 2017. What the stakeholder needs the ability to build arrival rates based on many factors, such as fleet mix, braking action, and other factors, and be able to accommodate limitations caused by construction or other environmental impacts. Being able to adjust AAR in advance for weather impacts would also be important in planning NAS performance. The Airport Decision and Support Tool (ADEST) is one tool that addresses these kinds of variables. ADEST builds an arrival rate to an airport based the airport's configuration generated by the TAF and a unique set of business rules specific for that airport. ADEST can generate the arrival rate as far in advance as the TAF applies (24- or 30- hours). The goal of ADEST is to generate arrival rates at 34 of the busiest airports in the National Airspace System.

Ground Delay Program (GDP) Analysis Capability Tool

The GDP Analysis Capability Tool is a new addition to a suite of tools available to the PERTI team. The GDP Analysis Capability Tool allows traffic managers to search historical ground delay program data, compare it to similar historical constraints and use that information to predict future NAS performance.

The **TAF** is the weather forecast that generates the airport's configuration, in turn, **ADEST** predicts arrival configuration and rate that can be verified and compared to historical data provided by the **GDP Analysis Capability Tool** and **ASPM**.

PART 3 EXISTING PHILOSOPHIES

Collaborative Decision Making (CDM)

Collaborative Decision Making (CDM) defines an operating philosophy where Air Traffic Flow Management (ATFM) decisions are based on a shared, common view of the National Airspace System and an awareness of the consequences these decisions may have on the system and its stakeholders. There are two central tenants to CDM; that better information will lead to better decision-making, and tools and procedures need to be in place to enable air navigation service providers and the flight operators to more efficiently respond to changing conditions. By sharing information, values, and preferences, stakeholders learn from each other and build a common pool of knowledge, resulting in ATM decisions and actions that are most valuable to the system (CDM, 2017).

Collaborative Trajectory Options Program (CTOP)

Collaborative Trajectory Options Program (CTOP) introduced the concept of stakeholder being able to identify and communicate their preferences and criteria. Giving aircraft operators greater flexibility with the ability to manage their schedule in regards to both route assignment and delay as defined by *their* goals and objectives.

Plan, Execute, Review, Train, and Improve (PERTI)

Plan, Execute, Review, Train, and Improve (PERTI) defines the philosophy that focuses on the organizational processes and operational procedures that drive efficiency. PERTI describes operational roles and processes that will enable the integration of analytics into training, operational planning, and execution.

CDM defines the systemic, information sharing philosophy. **CTOP** introduced a concept how the CDM theory can be implemented electronically as the stakeholders identify *their* NAS management preferences and criteria. While **PERTI** describes the organizational processes and operational procedures that apply the CDM philosophy to meet the needs and goals of the stakeholders.

PART 4 A NEW PARADIGM

A paradigm is defined as a new way of doing things that have a huge effect on business or industry. With the emerging concepts of online collaboration, e.g. Decision Lens, the implications of Collaborative Decision Making (CDM) has limitless and endless possibilities for advance planning, especially within the PERTI process. Some of the most powerful tools for online collaboration are applications within the visual analytics tool set.

As discussed, CTOP introduced the concept of the stakeholders defining *their* preferences and criteria into the development of traffic management initiatives. The purpose of this white paper is to present the concept into a systemic CDM practice by way of online collaboration utilizing visual analytics.

In today's traffic management environment (1) the system identifies a constraint, (2) develops a goal based on the constraint, (3) implements an initiative to manage the identified constraint or goal. The PERTI process will close the loop by (4) reviewing the implemented initiative, (5) determine best practices and thereby, (6) reintroduce the best practices into the system to develop a learning culture. This concept is based on analytical thinking and not systems thinking (Senge, 2006).

The paradigm shift will occur when we, as a system, can: (1) prioritize the stakeholder's NAS management preferences (goals), (2) evaluate various courses of action for better results, (3) define decision criteria identified and possible choices, (4) determine the stakeholders that are relevant to the decision-making process, (5) and by using the criteria that have been established, the stakeholder access each component of the decision and decides what is important to *them.* This paradigm would enable the NAS to focus on *what needs to be done* rather than *how it is done*.

The paradigm shift is that the process occurs prior to identifying the constraint. As stated, what is needed is for the stakeholders is to insert *their* NAS management goals (or criteria) into the center of the decision-making process, enabling us (the NAS) to prioritize which initiatives or ideas are the best way to meet the defined needs and goals. Then apply the decisions to adapt to the constraint to manage the National Airspace System.

The paradigm does not change the role of the air traffic facility, which is the safe, orderly, and expeditious flow of air traffic. In addition, the paradigm does not change the role of the Air Traffic Control System Command Center, which is to manage the NAS. However, it does provide them with the information to enhance efficiency and customer service to the level that is demanded by the flying public.

Senge, P. M. (2006). *The fifth discipline: The art and practice of the learning organization*. New York: Doubleday/Currency.

PART 5 WE HAVE THE DATA!

Many factors influence a stakeholder's behavior and their preferences. The factors that affect behavior and preferences of stakeholders can change daily. For example, JetBlue earns 70% of their profits between June 1st and Labor Day. However, as traffic managers, we manage convective activity within the National Airspace System utilizing the same principles and criteria on May 21th as we do on July 10th as we do on September 8th. Do you think that JetBlue views those three days differently? Therefore, the question is, what are some of the factors and/or variables that influence stakeholders' behavior and criteria. Sorry, that was a poorly worded question; the answer is, of course, money and/or profits. The real question should have been, what are the variables within the control of the stakeholder or the FAA that can be managed or compensated for to maximize stakeholder goals? In other words, what are the variables that concern stakeholders? Sorry, that was another poorly worded question, because the stakeholders do not create money, they generate a profit by not losing revenue. Therefore, the proper question should be, what are the variables that the stakeholders and the FAA can manage or compensate for that concerns the customers of the stakeholders? The answer to that question ***must*** be the goals and criteria that we (the traffic managers of the NAS) base our decisions on and ***what needs to be done*** to manage the NAS.

The data is out there. In today's data-rich environment, if the data is not available, then it is not important. The question is where is the data? That is not that hard of a question, the hard question is how we can utilize the data and make the information from the data available to the people who are making the decisions. Because we know what information we want. The customers are concerned about departing on time (on time departure (D-0)), customer are concerned about arriving on time (arrival within 14 minutes (A-14)), customer are concerned about completion factors (because they are spending money to get from point A to point B), they care about accurate information and how long their flight is delayed. However, it is the service provider and/or the stakeholder that *must* identify the criteria to ensure that (1) the criteria, preferences, and goals are properly defined and (2) that those criteria, preferences, and goals are managed, and meeting expectations.

We, as a system, have the data!

PART 6 How Do We Do It?

How do we do it? Wow! That is an intimidating question, or it? I have tried to define all of the pieces of the puzzle, however, let us try to assemble the pieces to form a picture (or a vision).

What do we have? We have a **NEED**. We have **Existing Technologies**, as in the shape of the **TAF**, **ADEST**, **ASPM**, and the **GDP Analysis Capability Tool**. Traffic managers have defined and accepted philosophies, as in **CDM**, the concept introduced by **CTOP** where the stakeholders can identify *their* preferences and criteria to influence traffic management initiatives, additionally, the philosophy of **PERTI** that defines the organizational processes and operational procedures that implement these concepts and beliefs.

Most of all, we have explained the paradigm shift of identifying the stakeholder's NAS management criteria, preferences, and goals before determining a constraint, by utilizing online collaboration and visual analytics.

As the NAS enters into a situation (a weather scenario, a holiday event, or a peak of robust revenue potential), the stakeholders have defined *their* criteria and goals accordingly. This situation occurs in today's environment and is apparent by watching cancellation trends during the middle of the winter snow event compared to a holiday weekend during the summer as a convective weather event unfolds. However, as traffic managers, we are trained on how to manage the NAS. The point is, traffic managers need to consider *what needs to be done* rather than *how it is done*.

So, how do we do it? Simple, we ask what is relevant to the stakeholder (or the customers of the stakeholder) and then base the decisions on *their* NAS management criteria. The first question that comes to mind is why doesn't the FAA ask the customers directly what is important, simple, that is not the FAA's job. That is the job of the marketing departments of the stakeholders. Our job is to meet the *needs* of the stakeholders of the NAS. However, more importantly, it is our duty to listen to the *needs* and expectations of the stakeholders, just as it is the stakeholder's responsibility to hear the needs and expectations of their customers.

This is where the online collaboration utilizing visual analytics are introduced (see figure 1). Envision a communication conduit (website) where the stakeholder could define their business priorities (NAS management criteria, preferences, and goals) by the simple drag of a mouse. Picture a square with a dot in the middle of it, the four corners representing four different defined criteria, e.g. completion rate, predictability, the amount of delay received, and on-time departures (it could be any four different criteria, it could be eight different criteria by changing the shape of the conduit). Now, the stakeholders have the ability to drag the dot to indicate their business model, their company goals, their preferences or criteria for the week, the day or the shift.

Once the stakeholders have defined *their* NAS management criteria, the GDP Analysis Capability Tool powered by ADEST, and ASPM powered by ADEST can provide the stakeholders with the results of *their* criteria (based on the forecast, the airport's predicted arrival configuration, and arrival rate, along with historical performance data during similar weather conditions) for the purpose of evaluation. For example, if United Airlines states that *their* NAS management preferences or criteria for EWR on a specific day are to limit diversions and concentrate on completion factors, ADEST would determine the configuration and arrival rates and provide that information to the GDP Analysis Capability Tool and ASPM. In turn, the communication conduit would provide United Airlines with historical data of similar scenarios that meet their NAS management criteria of minimal diversions and high completion rates. In other words, if a stakeholder, on a holiday weekend, who defined completion rate as *their* most important criteria, would they accept (or not accept) a three-hour delay into New York in fulfilling their customer's expectations. However, if a stakeholder determines that there is a limited amount of acceptable delay before they start canceling flights, that information is provided. That information would be available to the decision makers. That information could look like, 12 out of 14 times that the decision was made to utilize a higher arrival rate in an initiative, the results were, that on average, the system experienced 35 more aircraft diversions.

The goals are to:

- Eliminate the obstacles to enhance clear decision making.
- Identify which decisions would help reach the defined goals.
- To be able to look back at a situation and not to say – If I would have known that, I would have decided differently.
- To accomplish all of this in a transparent and collaborative environment.

This concept allows the construction of realistic shift goals based on real data, defined by the stakeholders based on their expectations. In other words, if a stakeholder's objectives are to concentrate on passenger connections and *they* decide that they would like to delay their west coast departures to favor their east coast departures. The stakeholder might mitigate the risk by adding additional fuel to their aircraft to allow for airborne holding, then why should *our* goals be to try to limit diversions? *Their* objective was for their east coast passenger connection and planned for the possibility of diversions. With the power of online collaboration utilizing visual analytics, the stakeholders can define *their* NAS management preferences and criteria based on real data, which would allow the development of realistic goals that have real meaning. An example of a realist goal based on data could be: to be at least 95% compliant of managing EWR airport's goal of fewer than 18 diversions and a completion factor of greater than 93%. The goals were collaboratively developed by the stakeholders defining *their* criteria of limiting diversions and concentrate on completion factors, and by utilizing the stakeholders' criteria. Also, the goals were developed using real data (18 diversions as provided by ASPM powered by ADEST and a completion rate of 93% as provided by the stakeholder).

The realistic operational goals (PLAN) translates into operational practices (EXECUTE) that are implemented based on real data that can be quantified based on historical data (REVIEW) and introduced into a learning culture (TRAIN). The improvement part of the puzzle will mature on its own.

The paradigm shift defined in this white paper is based on identifying the stakeholder's NAS management criteria, preferences, and goals prior to determining a constraint. However, the NAS can utilize the same paradigm shift during other situations, e.g. security events. As an example, during the recent security event at Fort Lauderdale International Airport where there was an active shooter in the terminal, the goal under the new paradigm would have been based on securing the airport and decision would be built on that goal. Again, the paradigm centers on *what needs to be done* rather than *how it is done*. Once the paradigm is in place, and the technique of establishing new goals implemented, everyone is now working toward *what needs to be done* instead of different groups of people just doing things.

Figure 1: Example of online collaboration utilizing visual analytics

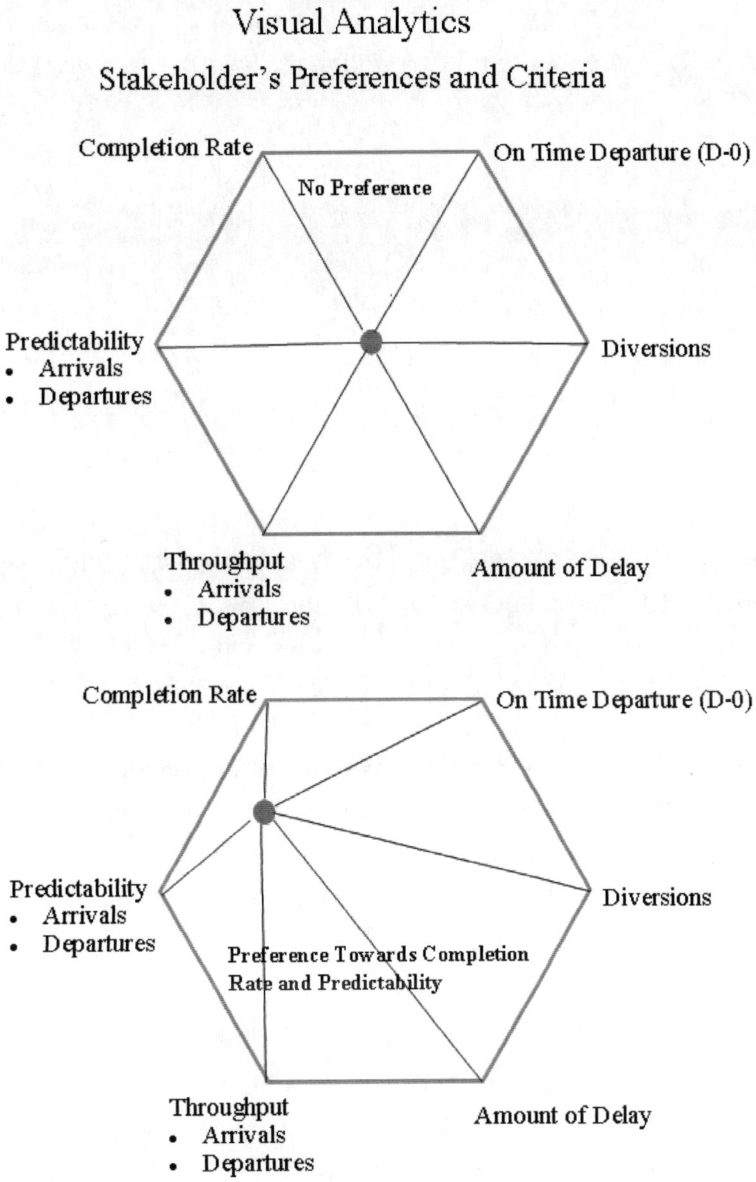

CONCLUSION

There is a fable about an interesting experiment that started with five monkeys in a cage. A banana hung inside the cage with a set of steps placed underneath it. After a while, a monkey went to the steps and started to climb towards the banana, but when he touched the steps, he set off a spray that soaked all the other monkeys with cold water. Another monkey tried to reach the banana with the same result. It did not take long for the monkeys to learn that the best way to stay dry was to prevent any monkey from attempting to reach the banana.

The next stage of the experiment was to remove the spray from the cage and to replace one of the monkeys with a new one. Of course, the new monkey saw the banana and went over to climb the steps. To his horror, the other monkeys attacked him. After another attempt, he learned that if he touched the steps, he would be assaulted.

Next, another one of the original five was replaced with a new monkey. The newcomer went to the steps and was attacked. The previous newcomer joined in the attack with enthusiasm!

Then, a third monkey was replaced with a new one and then a fourth. Every time a newcomer approached the steps, he was attacked. Most of the monkeys beating him had no idea why they were not allowed to climb the steps or why they were joining in the beating of the newest monkey.

After replacing the fifth monkey, none of the monkeys had ever been sprayed with water. Still, no monkey ever approached the steps. Why not? Because as far as they knew, it was the way it had always been done around here.

The point is that sometimes we as a system do things because that is the way it has always been done, like identifying a constraint before defining a goal. The purpose in writing this white paper was to introduce just enough fact, theory and philosophy to inspire and encourage change. Recommendations for further research is needed to define visual analytics in an online collaboration environment, in addition, development needs to be accomplished to build a communication conduit from ADEST to the GDP Analysis Capability Tool and ASPM (i.e. system to systems (SoS) development). Without a willingness to change and a sensible vision; we will continue to do the same thing and sometimes not even know why. However, there is one thing that will always remain; the National Airspace System will continue to define how the world manages the ever-changing environment that the flying public, of the globe, demands and deserves.

Thank you,

Jeff Burke

Remote Identification of Small UAS

Jared Ablon, Gregory S. McNeal, & Stephen Willer
AirMap

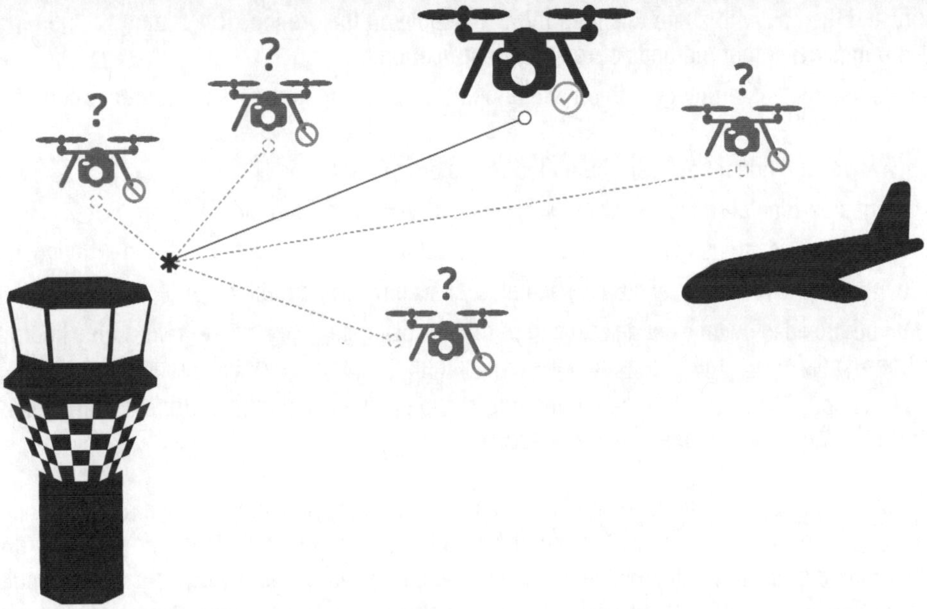

AIRMAP

Introduction

Today, there is no widely accepted, reliable method of remotely identifying a drone as it flies overhead - a fundamental barrier to safely integrating drones into the national airspace.

Identification is a critical part of any air traffic control system – whether that traffic is manned or unmanned. When aircraft identities can be verified and trusted, controllers can make traffic management decisions with confidence, ensuring safe skies for all. But if their identities cannot be verified, uncertainty and risk are introduced into the system, affecting the safety and security of the airspace.

With the advent of ADS-B and the development of NextGen, the world of aviation is moving towards digital methods of identification and communication - and enduring the growing pains of introducing new technologies into an existing system. Drones, however, are just beginning to take flight, presenting the opportunity to build secure, authenticated methods of remote identification into the system from the very beginning.

Typically, the solutions proposed to identify drones from the ground require the addition of new, costly, or heavy hardware to drones or the use of proprietary software, which could discourage wide adoption and place an undue burden on the drone industry. Fortunately, most drones are already equipped with technology that could be used to enable remote identification by authorities on the ground.

This paper recommends a technology solution (SSL/TLS certificates) that provides both secure authentication of a drone's identity and the ability to share this identity with people on the ground. This method comes with several important advantages over other methods of remote identification:

- It can be used to authenticate with an unmanned traffic management (UTM) system integrated with existing ATC.
- It requires no additional hardware be added to the drone.
- It leverages existing broadcast capabilities, such as WiFi, Bluetooth, or LTE.
- It allows authorities to use existing smartphone technology to receive a drone's identifying information.
- It is low-cost, highly secure, and easily scalable to accommodate millions of drones.
- It could be provided by an existing and competitive ecosystem of certificate authorities (more than 1480 identified to date), instead of requiring a single company's proprietary technology.
- It prevents spoofing/hacking of identifying information and ensures that a drone's identity can be trusted by air traffic controllers and others in the airspace.

If SSL/TLS digital certificates, coupled with existing broadcast technologies and a mobile application, are used for remote identification, manufacturers would be compliant with just a firmware upgrade; no costly hardware adjustments to drone designs or production would be required; personally identifying information would be closely protected; and authorities would be able to effectively identify drones and their operators when required. Critically, air traffic controllers would be able to trust that each drone sharing digital information with the national airspace system could be authenticated.

This paper outlines the key features of this proposed method.

Assumptions

1. Air traffic controllers would receive information about a drone's location through the UTM system.

2. Air traffic controllers would be able to learn whether that drone's identity had been verified and its identifying number – but not view the personally identifying information of the operator.

3. The public should also be able to use their smartphones to see this information. The identifying number can be used by members of the public to report an issue to authorities, just as they might report the license plate of a speeding car on the roadway, or the N number of a registered manned aircraft. Authorities can also use the identifying number to associate the aircraft with its FAA registration number. If required, this method can be used instead to ensure that only authorities have access to the drone's identifying number. In that case, additional security and privacy controls would be added to restrict access to this information.

4. The public will receive general information about the location of the drone pilot (city, state, country). This information can be appropriately redacted to ensure that pilot information is protected.

5. The use of SSL/TLS certificates for remote identification also has a cybersecurity benefit, which can enable encryption, digital signatures, and mutual authentication of data shared by drones - thereby bringing another layer of security to remote identification of drone operators.

Technology Description

SSL/TLS certificates (x509 certificates) will work with most existing drones without the addition of new hardware. Activation for widespread compliance will require only a firmware/software update.

SSL/TLS certificates are a digital technology commonly used to secure communications on the internet and in the Internet of Things (IoT) ecosystem. Each SSL/TLS certificate establishes a pair of digital "keys" that are used to encrypt and/or digitally sign information shared with others. On the internet, this is information shared between websites and their users (for example, online banking sites). When an individual sends information to a website with https, it is encrypted with a public key, designated by a lock next to web link in the address bar. Only the web page visited has the private key needed to decipher the message. These keys can be used to encrypt and digitally sign information, reducing the risk of threats to information security, like hacking or spoofing, and are currently used in massive volumes for IoT and internet purposes. Certificates for drones will work on a similarly high scale.

Because drones cannot always be sensed by radar and other existing technologies, they will need to share their locations and intent by exchanging data within the UTM system. But how can that information be verified? How can air traffic controllers confirm that information about the "drone" they see on their user interface can be trusted?

Figure 1: SSL/TLS Certificates to Guard against Spoofing

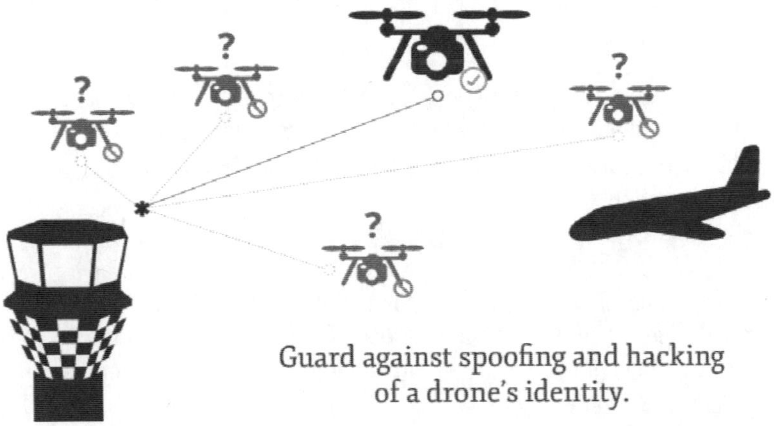

If a drone broadcasts information about itself or its flight without a digital signature, that information is unverified, and the recipient cannot tell if this information had been modified or spoofed. But if this information is "digitally signed" by an SSL/TLS certificate, the authenticity and integrity of the message can be verified, and it can be confirmed as belonging to a specific drone. When an air traffic controller receives information digitally signed by a drone, he or she knows that the drone's identity has been authenticated. The SSL/TLS certificate helps to ensure that the drone's identity has not been spoofed or hacked.

These certificates could also be used to enable three steps of remote identification by authorities:

- Verification of the drone's identifying number;
- Broadcasting of that identifying number to a user on the ground; and
- Secure access to that identifying number through a mobile app on a smartphone or tablet.

Figure 2: Remote Identification Process with SSL/TLS Certificates

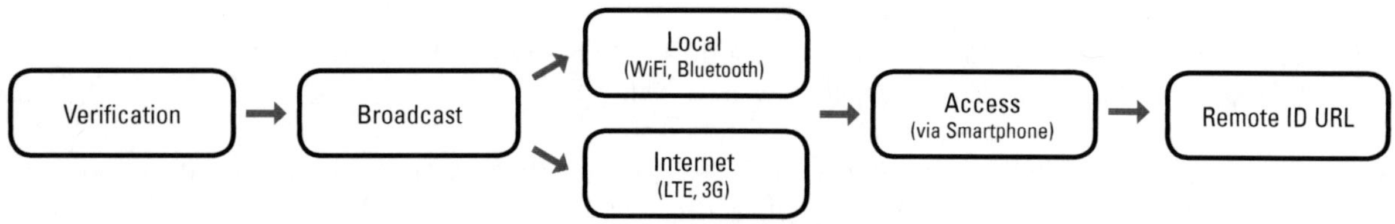

1. Verification

A Certificate Authority validates the drone operator's email address, phone number, name, and address and issues a x509 certificate SSL/TLS certificate. The certificate includes a unique identifying number for the drone, akin to a car's license plate, and a Remote ID URL, where authorities can learn how to access more detailed information in the case of an investigation.

2. Broadcast

To ensure compatibility with current technology – and leave room for future innovation – a performance standard, minimum broadcast distance (e.g. 1500 feet), should be determined without requiring a specific broadcast medium.

That minimum broadcast distance can be met using technologies already on board most drones. Most drones are equipped with WiFi, and with just a firmware upgrade, they can support WiFi Aware, a standard from the WiFi Alliance. It is a real-time, energy-efficient extension of WiFi that allows a device to remain discoverable by devices that may not have a GPS, cellular or hotspot connection. Similar discoverability or device-to-device communication is also possible using other standards, such as Bluetooth Smart (although longer broadcast distances may be more difficult to achieve).

As a complement to local broadcast, an internet connection, such as LTE, could also be used. However, the drone should also be equipped with a local broadcast so that the drone's identifying number can be shared with recipients without an internet connection. This ensures that remote identification is still available in remote areas and areas with limited or no cellular coverage.

3. Access

Authorities and others on the ground view the drone's ID number and Remote ID URL via a mobile app on their smartphone, tablet, or other device. The drone's position is visualized on a map; users tap to view the drone's ID number and Remote ID URL. Members of the public can use the drone's ID number to report issues to authorities, but cannot access personally identifying operator details.

Figure 3: Accessing Drone Identifier & Remote ID URL

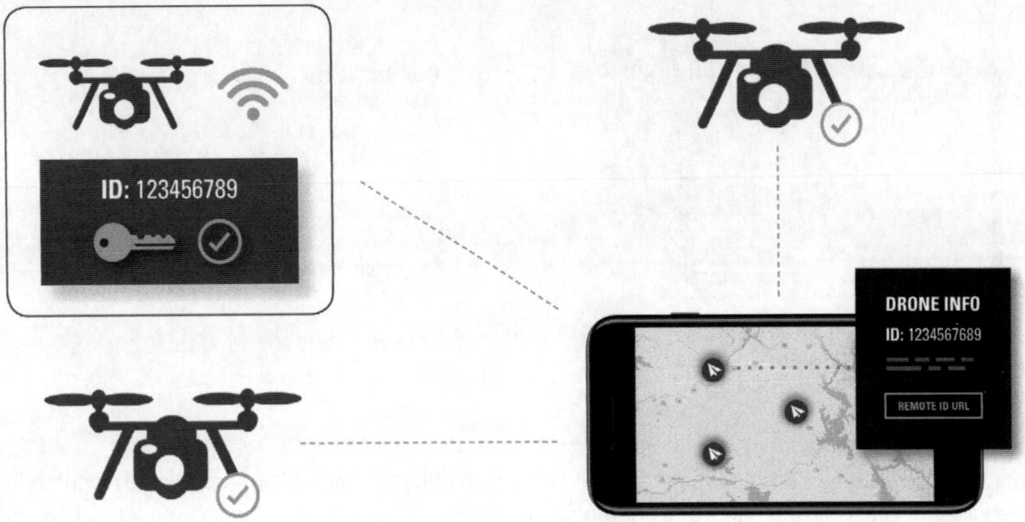

The mobile application should be able to receive information via WiFi, Bluetooth, or the internet (e.g. LTE), and serve that information to users. This allows for maximum flexibility for manufacturers, who could broadcast local information from their drones in the broadcast medium of their choice. The receiving app would then relay this information to a cloud solution that would redistribute it to other app users, such as a law enforcement officer standing just outside the minimum broadcast range. Only one person would have to be within the local broadcast range of the drone for others to also be able to see the unique identifier information.

Technology Operational Concept

User Interface Application

There will be a publicly available application available on mobile devices for users to download. When an app user sees a drone flying nearby, the user can pull up the app and retrieve the authenticated identification that is locally broadcast or shared via the internet by the drone.

This application will include a map, with the location of the drone corresponding to the drone's position in the sky. When the user taps on the drone in the application, they will receive information stored in the TLS certificate (drone identification number; city, state, country; and a Remote ID URL). This information can be further restricted to just the identifying number and/or the Remote ID URL if required.

The drone operator could opt in to provide additional information on the Remote ID URL, or restrict this information to just their drone identifier. In addition, law enforcement and regulatory authorities can visit the Remote ID URL to learn how to access additional information about the drone operator (name, phone number, email address, address) under restricted conditions (such as in the case of an investigation). This information would only be available to authorities and is not accessible to the general public. All publicly identifiable information (PII) would be encrypted at rest and in transit.

Figure 3: Example Remote ID URL

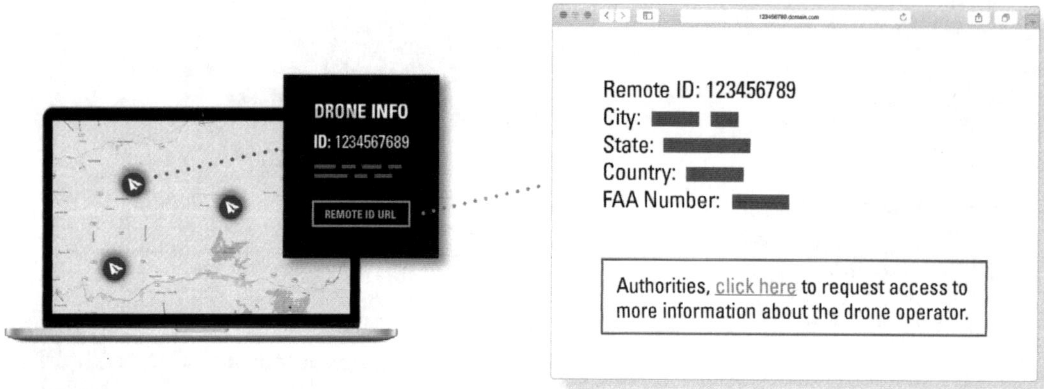

Supporting Infrastructure

This method of Remote ID will ensure the greatest amount of compliance and easiest adoption by industry because the main requirements are technologies that are already integrated with most existing drones: a local broadcast technology, such as WiFi or Bluetooth, and ideally, but not necessarily, a cellular technology such as LTE. Using these technologies ensures the widest dissemination of data without any custom hardware needed for the drone (e.g., additional broadcast or receiver hardware, such an ADS-B receiver for people on the ground).

However, the proposed solution can also be employed to enable authentication of information broadcast via ADS-B. One of the benefits of SSL/TLS certificates is that they enable secure and verified identification, while remaining agnostic as to the broadcast or receiver technologies. Thus, this solution can enhance the security of ADS-B if that hardware mandate is deemed to be an additionally acceptable broadcast method.

Root certificate authorities are also supporting infrastructure for this solution, with more than 1480 potential providers already in operation. There is substantial competition in this space, and a vast infrastructure already exists to support this method of remote identification. Certificates are a proven solution for authenticating and securing financial and other transactions on the internet, and can be leveraged to similarly secure and authenticate drones.

Technology Readiness Level

All of the technologies required by this proposal are in production today, with a Technology Readiness Level of 9. Software development is needed to build out the solution, but no research is needed because all technologies are already proven and widely understood by software developers and IoT companies.

Reliability Assurance and Continuity of Service Features

Since the system works locally via WiFi or Bluetooth, it can share identifying information with any user within range. Additionally, the system could be enhanced by a backend server that redistributes data received by a single user via local broadcast to those outside the broadcast range. Backend servers can be made available at scale so that the system is robust, whether it receives information locally or via the internet. There can also be multiple registrars, so that data is not hosted by a single entity.

Key Questions

How is privacy protected?

This Remote ID implementation only broadcasts an ID number and optionally, city/state/country information for the drone. It includes no Personally Identifiable Information (PII). While PII can be obtained by credentialed authorities in the event of an investigation or other conditions determined by regulators, this information is not available to the general public. If the drone operator would like to share identifying information on the Remote ID URL, they can opt in to do this - but it is not required and not the default.

Why use a digital certificate rather than a physical number affixed to the aircraft?
Drones communicate digitally and fly hundreds of feet in the air, which means that a physical identifying number cannot be identified from the ground. Using SSL/TLS certificates will allow authorities to access identifying information under appropriate circumstances, while ensuring the privacy of operators.

Who will provide the certificates? Are certificates expensive?
A wide range of providers from the existing certificate industry could offer this service to the drone industry. Certificates are standard practice on the internet and in the Internet of Things industry, with both free and paid options available. For example, Amazon Web Services' IoT service, Microsoft Azure's IoT hub, and root certificate authorities such as GlobalSign and DigiCert all offer certificates for other IoT devices. We expect that these and many other root certificate authorities will provide these services to drones. Paid certificates for internet websites are typically $20-$100 for one year.

The market has many players and is very mature. There are currently 180 root certificate authorities and hundreds of additional intermediate certificate authorities that are trusted by Mozilla Firefox to issue certificates. In total, there are more than 1480 certificate authorities already in existence that could participate in providing this service to drones.

Are SSL/TLS certificates scalable and easy to use?
Digital certificates offer a scalable way to identify millions of drones, just like they offer a scalable way to identify entities that own websites on the worldwide web. With modern hardware, proper tuned TLS deployment, and mechanisms like TLS resumption and TLS false start, the overhead and added latency of the security protocol is minimal.

Conclusion

Any method selected to remotely identify drones must leverage existing technology and broadcast capabilities; prioritize security for the privacy and protection of identifying information; and be easy to implement and use for manufacturers, developers, operators, authorities, and air traffic controllers. And in order to ensure compliance, the selected method cannot mandate the addition of new or expensive hardware.

SSL/TLS certificates, broadcast via existing technologies (WiFi, Bluetooth, and LTE) and accessible via a smartphone app for the public, fully satisfy these requirements, providing a secure method to identify drones, and when required, allowing authorities to identify operators of unmanned aircraft. The result: a safer, more secure, and more accountable drone ecosystem.

The Effects of Severity of Losses of Well Clear on Minimum Operations Performance Standards End-to-End Verification and Validation Simulation Study for Integrating Unmanned Aircraft Systems into the National Airspace System using Detect and Avoid Systems

Rania W. Ghatas
NASA Langley Research Center, Hampton, Virginia 23681

Vincent P. Schultz
NASA Langley Research Center, Hampton, Virginia 23681

Devin P. Jack
Adaptive Aerospace Group, Hampton, Virginia 23513

Dimitrios Tsakpinis and James L. Sturdy
SAIC, Inc., Hampton, Virginia 23681

Michael J. Vincent
NASA Langley Research Center, Hampton, Virginia 23681

Keith D. Hoffler
Adaptive Aerospace Group, Hampton, Virginia 23513

Robert Myer
SAIC, Inc., Hampton, Virginia 23681

Anna DeHaven
Craig Technologies, Inc., Hampton, Virginia 23681

I. Introduction

AS Unmanned Aircraft Systems (UAS) make their way to mainstream aviation operations within the National Airspace System (NAS), research efforts are underway to develop a safe and effective environment for their integration into the NAS. Detect and Avoid (DAA) systems are required to account for the lack of "eyes in the sky" due to having no human on-board the aircraft. The current NAS relies on pilot's vigilance and judgement to remain Well Clear (CFR 14 91.113) of other aircraft. RTCA SC-228 has defined DAA Well Clear (DWC) to provide a quantified Well Clear volume to allow systems to be designed and measured against. Extended research efforts have been conducted to understand and quantify system requirements needed to support a UAS pilot's need to remain well clear of other aircraft. The efforts have included developing and testing sensor, algorithm, alerting, and display requirements. More recently, evaluation of sensor uncertainty and uncertainty mitigation strategies have been evaluated.

This paper discusses Severity of Losses of Well Clear (SLoWC) results and lessons learned from an end-to-end verification and validation (E2-V2) simulation study of a DAA system representative of RTCA SC-228's proposed Phase I DAA Minimum Operational Performance Standards (MOPS). SLoWC is a metric used "to assess the severity of Loss of DAA Well Clear on a per-encounter basis by capturing the most serious instance of Loss of Well Clear throughout an encounter" (RTCA, Inc., In Press). It is based on the severity of violation into all three of the Well Clear components, which include: Horizontal Proximity, Horizontal Miss Distance Projection, and Vertical Separation. The SLoWC range is from 0% (no loss of well clear) to 100% (mid-air collision). NASA Langley Research Center (LaRC) was called upon to develop a system that evaluates a specific set of encounters, in a variety of geometries, with end-to-end DAA functionality including the use of sensor and tracker models, a sensor uncertainty mitigation model, DAA algorithmic guidance in both vertical and horizontal maneuvering, and a pilot model which attempts to steer the ownship aircraft well clear from intruder aircraft, having received collective input from the previous modules of the system. LaRC had a functioning batch simulation and added a sensor/tracker model from the FAA Tech Center, in-house developed sensor uncertainty mitigation strategy, and an in-house developed pilot model similar to one from MIT Lincoln Laboratory (MIT/LL). The resulting simulation provides the following key parameters, and many more, to evaluate the effectiveness of the MOPS DAA system: SLoWC, closest point of approach (CPA), alert scoring, and alert jitter. The technique, SLoWC results, and lessons learned from a detailed examination of DAA system performance over specific test vectors and encounter cases during the simulation experiment will be presented in the paper.

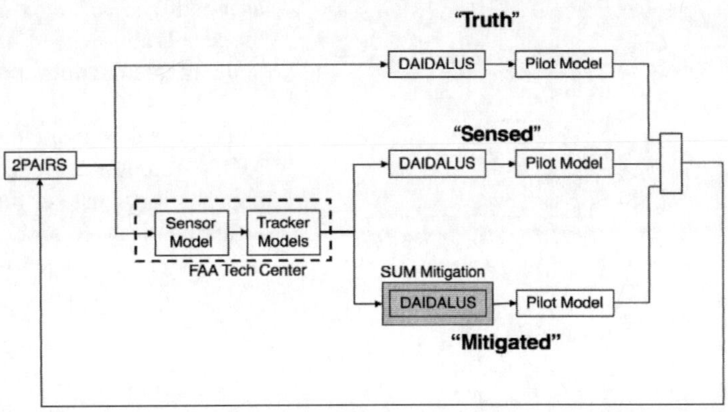

Figure 1. E2-V2 Simulation Architecture

II. Method

An end-to-end fast time simulation tool was developed that encompasses simplified unmanned aircraft (UA) maneuver dynamics and all of the components of a DAA system. The system provides for one UA and a single intruder to fly a pre-determined encounter trajectory while having the UA either continue the trajectory and measure DAA system alerting or follow DAA system maneuver guidance per a simple deterministic pilot model. Figure 1 shows the simulation architecture used in E2-V2, which portrays the data flow of the simulation used in the study including a model of aircraft dynamics, a representative DAA algorithm, a sensor/tracker model that encompasses uncertainty modeling, and a deterministic pilot model that was used to close the loop on aircraft encounters. Aircraft dynamics are modelled using the 2-degrees-of-freedom Prototyping Aircraft Interactions Research Simulation (2PAIRS) tool. DAIDALUS (Detect-and-AvoID Alerting Logic for Unmanned Systems) is the representative DAA algorithm that computes maneuver guidance based on the ownship and intruder(s) state information from the sensor/tracker models or the uncertainty mitigation.

The deterministic pilot model was developed and provided by MIT/LL; NASA LaRC developed and implemented a functionally representative version for this simulation. The model was explicitly constructed to handle single intruder cases only and avoidance maneuvers in the lateral dimension. The Java implementation of the MIT/LL Matlab model deviates slightly from the source to distinguish the state-machine that governs timing in the processing functions.

The Sensor and Tracker models were developed and delivered by the Federal Aviation Administration (FAA) William J. Hughes Technical Center in support of RTCA SC-228. The sensors used in this study include: Automatic Dependent Surveillance – Broadcast (ADS-B) In, Active Surveillance Transponder (AST), and an air-to-air RADAR; each sensor was tested individually. The simulation architecture allows for the capability of flying in three modes, including:

- Truth: Uses perfect state information,
- Sensed: Uses degraded state information from the sensor, and
- Mitigated: Uses sensor degraded state information with a sensor uncertainty mitigation (SUM) approach

The SUM approach, used in the Mitigated mode, creates phantom aircraft position and velocity based on estimated sensor uncertainty (Figure 2). Scaling factors were optimized to reduce frequency and severity of losses of well clear and to increase probability of accurate alerts and guidance. Jack, et al. (2017) is a closely related paper presents the mitigation approach, results, and lessons learned from the SUM simulation study.

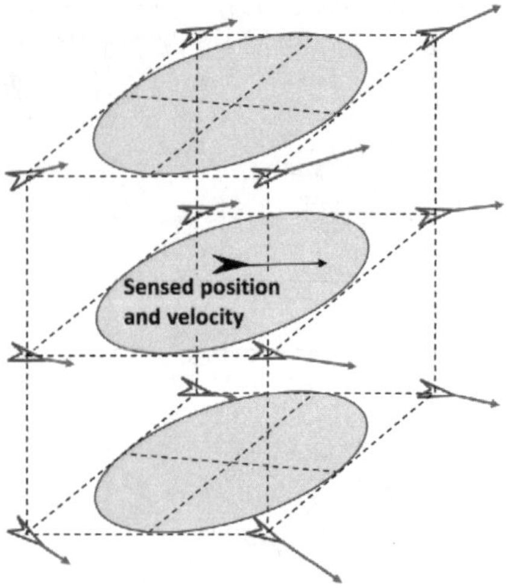

Figure 2. Sensor Uncertainty Mitigation Approach

Multiple encounters from multiple sources were designed to show, through detailed examination of specific test vector and encounter cases, whether a MOPS-representative DAA system behaves acceptably. Each encounter was run several times through all three modes (Truth, Sensed, and Mitigated) for replication purposes to verify and validate the output data.

The data was analyzed to determine the overall acceptability of a MOPS-representative system via the end-to-end simulation study. The paper will discuss, in further detail, the results of the study in addition to lessons learned and observations that were provided to SC-228 for use in developing the MOPS.

III. Scenarios

Multiple encounters were utilized to show whether a MOPS-representative DAA system behaved acceptability. Encounters were run in a closed loop simulation environment in order to mimic maneuvering behaviors of a human pilot in addition to human response delay times. A fixed pilot delay time, relative to alert issuance times, was used to make sensor uncertainty the only variable between runs of the same sensor/encounter set. Open loop encounters were also run, which provided the ability to characterize the original encounter geometry, with no pilot response, along with timing and alert jitter issues. Open loop encounters were compared to the closed loop data.

Encounters originated from two main categories: National Airpsace System (NAS)-Derived Encounter Sets and MOPS Requirements-Derived Test Vectors. This paper, however, focuses on the SLoWC results of the MOPS Requirements-Derived Test Vectors.

The MOPS requirements-derived test vectors will be included as supplement to the Phase I DAA MOPS. Each test vector, or track, was placed in one of two categories: alerted or non-alerted. For the E2-V2 simulation study, only alerted tracks were utilized for data collection. "Alerted tracks test the alerting capabilities of a DAA system for a range of aircraft encounters that have either occurred historically in the en-route environment, or have been identified through flight test or the design of prototype DAA systems to stress the performance of a DAA system" (RTCA, Inc., In Press).

The tracks were derived from multiple sources, including:

- A review of mid-air collisions that occurred between January 2000 and June 2010
- 95 Stressing Cases used by the Science and Research Panel (SARP) for the derivation of the DAAWC boundaries
- Flight Test 4 conducted by NASA in support of DAA MOPS development
- Test Vectors used in RTCA DO-317B for testing of the Airborne Surveillance and Separation Assurance Processing tracker and TSAA

Test vectors describe cases that are representative of encounters observed during routine operations and categorized as Head-On, Converging, Overtake, and Maneuvering encounters. Additionally, test vectors also described encounters that are considered to be "corner cases" that stressed the performance of the system, such as High Speed encounters (RTCA, Inc., In Press).

Table 1 shows the final closed and open loop encounter set used for each category; the final numbers are based on Truth tracks. Closed and open loop runs had identical encounter sets in each category for analysis comparison purposes. The column titled "Total" shows the initial number of test vectors developed. The remaining three columns show the number of encounters according to category description and sensor type after the initial encounters were filtered.

Table 1. Final E2-V2 Test Vectors Set

		E2-V2 Test Vectors (based on Truth Tracks)		
Category Description	**Total**	**Radar**	**AST**	**ADS-B**
MOPS: Head-On	15	3	14	15
MOPS: Converging	20	10	13	20
MOPS: High Speed	4	0	4	4
MOPS: Maneuvering	15	2	14	15
MOPS: Overtake	16	3	15	16
Total	70	18	60	70

IV. Severity of Loss of Well Clear Results

This paper will focus on the Severity of Loss of Well Clear (SLoWC) results. SLoWC ranges from 0% (no loss of well clear) to 100% (mid-air collision).

4.1 Radar SLoWC

Figures 3 through 6 show SLoWC histograms for the radar guidance source in a closed loop simulation using MOPS Requirement-Derived test vectors.

Results for Head-On test vectors (Figure 3) show that in cases using Truth closed loop data for guidance, the ownship was always able to remain well clear from other aircraft (0% SLoWC). However, 1/3 of those same cases would have resulted in a mid-air collision (MAC) or near mid-air collision (NMAC) without the DAA system and pilot maneuvers (Open Loop Truth, 90-100% SLoWC). SLoWC performance for closed loop Sensed radar guidance was similar to closed loop Mitigated guidance, both of which are not far from performance for Truth data.

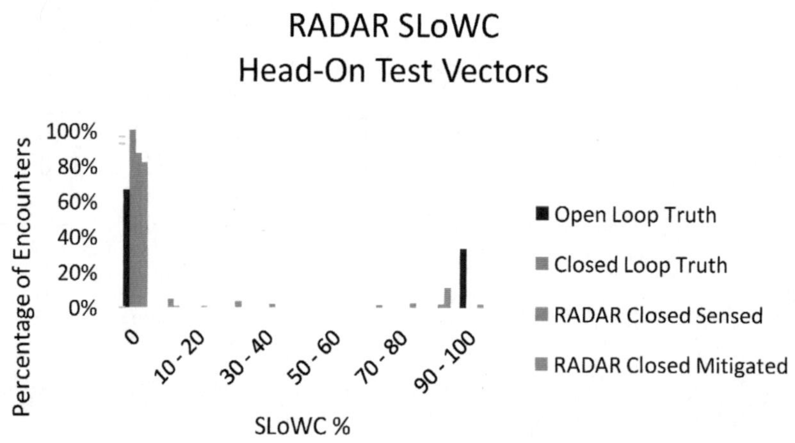

Figure 3. Radar SLoWC for Head-On MOPS Requirement-Derived Test Vectors

Similarly, results for Converging test vectors (Figure 4) show that in cases using closed loop Truth data for guidance, the ownship was always able to remain well clear from other aircraft (0% SLoWC). One-third of those same cases, though, would have resulted in a MAC or NMAC without the DAA system and pilot maneuvers (Open Loop Truth, 90-100% SLoWC). Also, as with the Head-On Test Vectors, for the Converging Test Vectors performance with Sensed and Mitigated guidance spread out the distribution somewhat, slightly reducing the number of encounters remaining well clear (0% SLoWC) and slightly increasing those with values between 0 and 60. Note that the DAA system was able to keep all Converging encounters using Sensed or Mitigated guidance below 60% SLoWC.

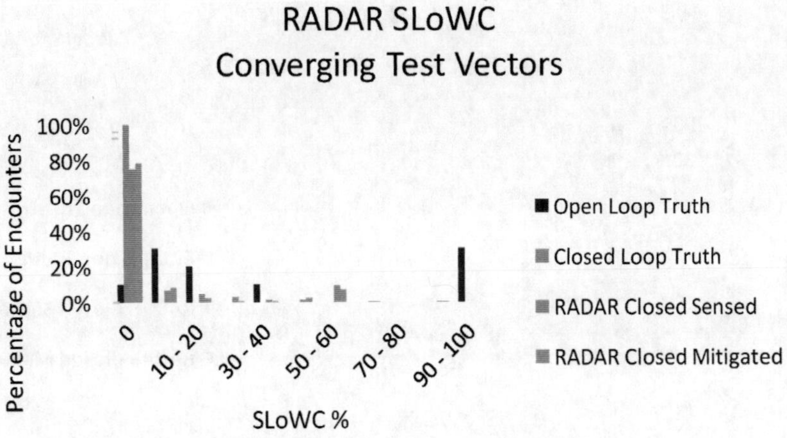

Figure 4. Radar SLoWC for Converging MOPS Requirement-Derived Test Vectors

Results for Maneuvering test vectors (Figure 5) show that the ownship lost DAAWC in all encounters, regardless of the source of guidance data, even Truth. Performance remained consistent across all four cases (open Truth, closed Truth, Sensed, and Mitigated) with 50% of encounters having a SLoWC between 0% and 10%. The other half of the closed loop Truth encounters resulted in a 30-40% SLoWC. Note that the other half of both the Sensed and Mitigated encounters resulted in lower severity scores (10%-20%), likely because the sensor uncertainty caused the pilot model to maneuver earlier than when using Truth guidance.

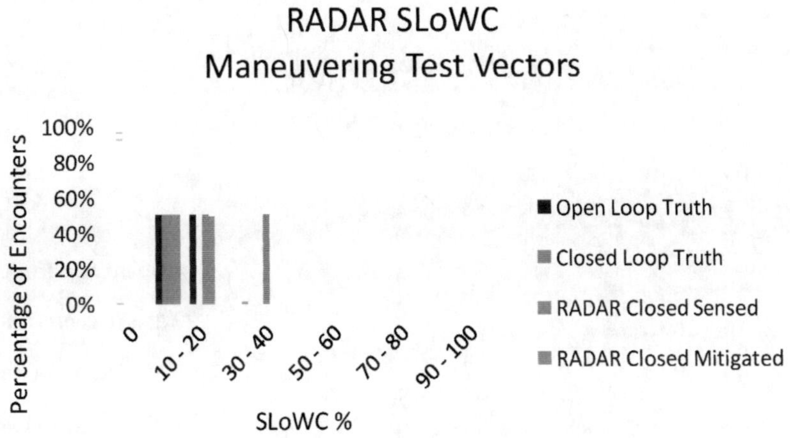

Figure 5. Radar SLoWC for Maneuvering MOPS Requirement-Derived Test Vectors

Figure 6 shows results for Overtake test vectors using radar data for guidance. The closed loop system resulted in 100% of encounters being within the 0% SLoWC range when Truth guidance was used. Open loop Truth results show 67% of encounters were within this SLoWC range and the remaining 33% of encounters were within the 20-30% SLoWC range. Overtake encounters using the radar sensor scored better in comparison to the other encounter geometries, which can be attributed to the radar's field of regard specifications enabling it to always detect the intruder because the ownship is overtaking, or approaching, the other aircraft from behind.

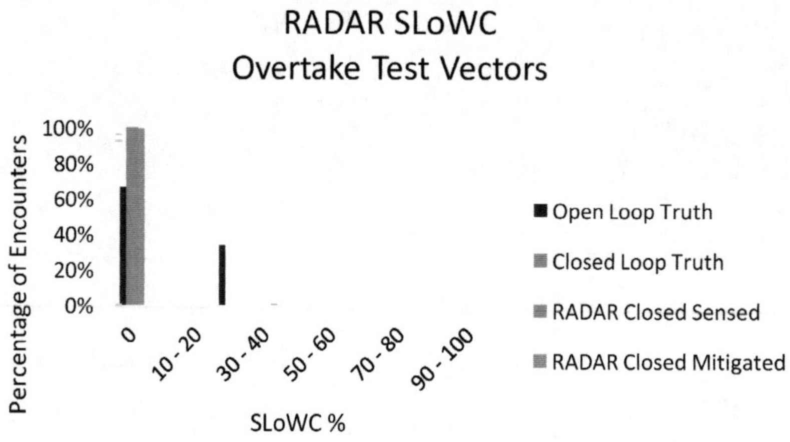

Figure 6. Radar SLoWC for Overtake MOPS Requirement-Derived Test Vectors

4.2 Active Surveillance Transponder (AST) SLoWC

Figures 7 through 11 show SLoWC histograms for the AST guidance source used in a closed loop simulation for MOPS Requirement-Derived test vectors.

Results for Head-On test vectors (Figure 7) show that the DAA system resulted in no losses of well clear for all encounters using Truth guidance. Sensed and Mitigated guidance encounters, however, were spread out across the entire range of SLoWC values, with performance similar to each other and to the open loop Truth results.

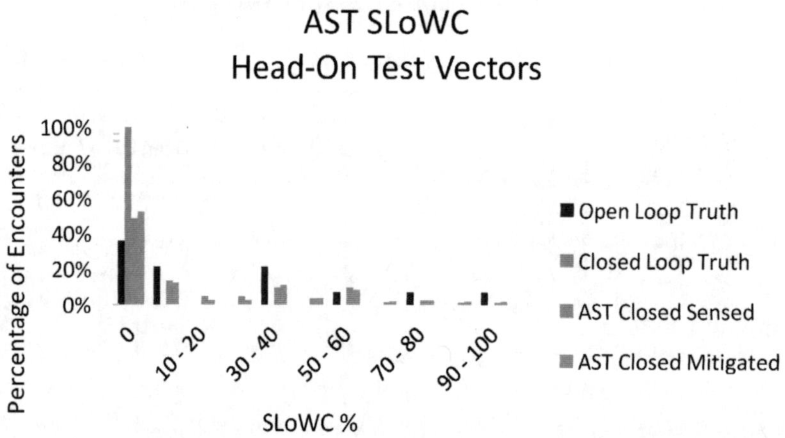

Figure 7. AST SLoWC for Head-On MOPS Requirement-Derived Test Vectors

Similarly, results for Converging test vectors (Figure 8) show that the DAA system resulted in 0% SLoWC for all encounters using Truth guidance. Sensed and Mitigated results were also spread out, but to only part of the SLoWC range. The SUM approach enabled the DAA system to pull the distribution to the left, eliminating all SLoWC values above 50% and increasing the number of well clear encounters to 79%.

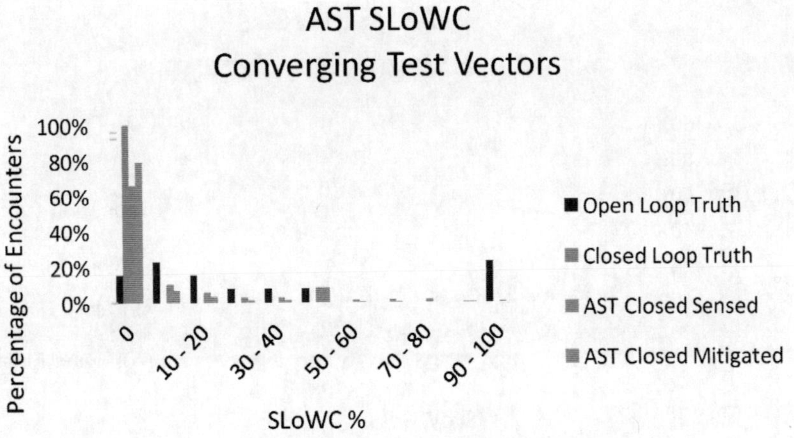

Figure 8. AST SLoWC for Converging MOPS Requirement-Derived Test Vectors

Figure 9 shows results for High Speed test vectors. Open loop Truth results show that all of these High Speed encounters would have some loss of well clear without any maneuvering to avoid it, including a significant portion (25%) having a SLoWC value close to 100%. Truth guidance in the closed loop system resulted in all encounters having 0% SLoWC. Both Sensed and Mitigated guidance spread the distribution out across essentially the entire range of SLoWC, but Mitigated guidance shows the SUM approach was able to keep 80% of the encounters well clear (SLoWC of 0%), and increase of 17% over Sensed guidance.

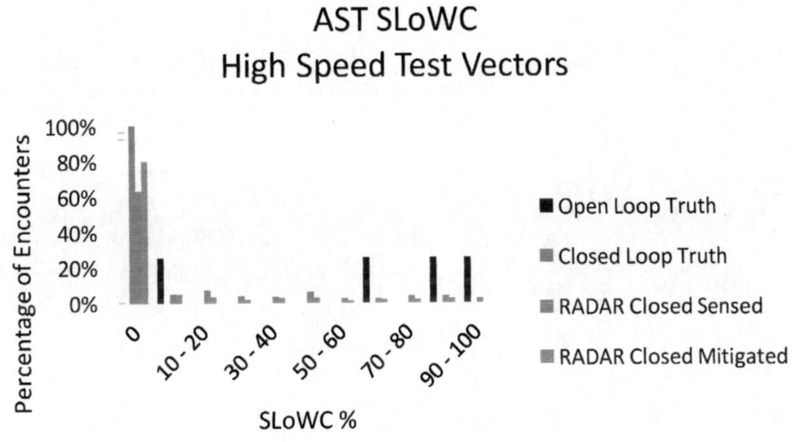

Figure 9. AST SLoWC for High-Speed MOPS Requirement-Derived Test Vectors

Figure 10 shows results for Maneuvering test vectors using the AST sensor. Open loop Truth results show that all of the encounters would result in some loss of well clear, and the DAA system was able to prevent loss of well clear in only small portions of all maneuvering (closed loop) cases. However, the Mitigated approach had the highest percentage of encounters with no loss of well clear (33.86%). Overall, the DAA system performed least favorably in the Maneuvering encounters using the AST sensor in comparison to the other encounter geometries and sensors.

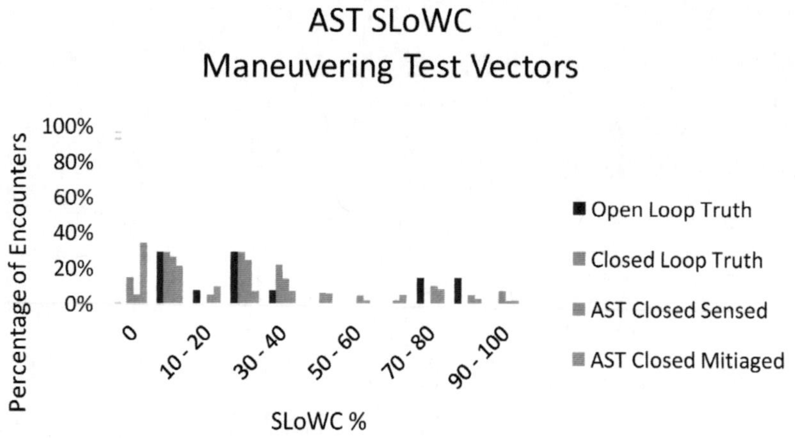

Figure 10. AST SLoWC for Maneuvering MOPS Requirement-Derived Test Vectors

Figure 11 shows results for Overtake test vectors. Similar to the radar Overtake results, the closed loop system resulted in 100% of encounters having a 0% SLoWC when Truth guidance was used, but with some distribution over the SLoWC range with no maneuvers (Open Loop Truth). Both Sensed and Mitigated guidance results are very similar, avoiding loss of well clear in most encounters.

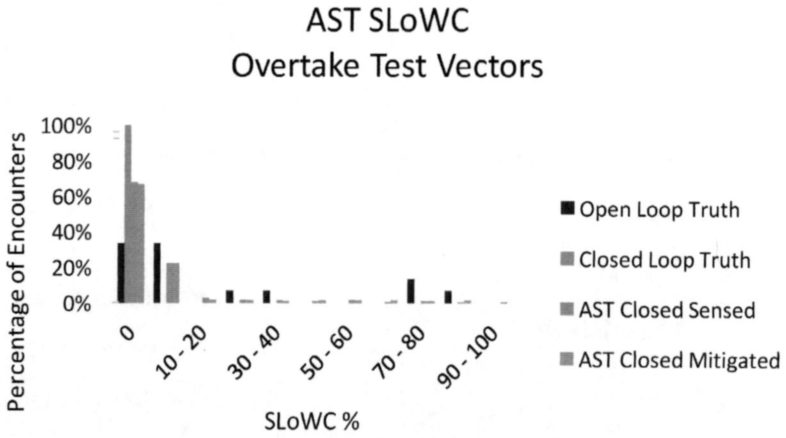

Figure 11. AST SLoWC for Overtake MOPS Requirement-Derived Test Vectors

4.3 Automatic Dependent Surveillance – Broadcast (ADS-B) SLoWC

Figures 12 through 16 show SLoWC histograms for ADS-B guidance source used in a closed loop simulation for MOPS Requirement-Derived test vectors. ADS-B data is by far less noisy and uncertain than radar and AST; the figures show that the DAA system performs very close to the way it would with perfect (Truth) data.

Results for Head-On test vectors (Figure 12) show that the DAA system resulted in no loss of well clear for all of the encounters using Truth guidance, whereas open loop Truth results show a 0% SLoWC value for only 33% of encounters. Essentially all the encounters using both Sensed and Mitigated guidance had a SLoWC of 0% (96% for closed loop Sensed and 97% for closed loop Mitigated).

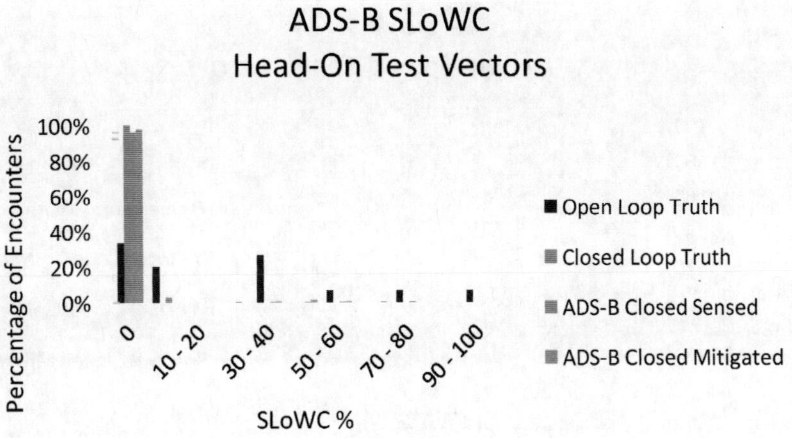

Figure 12. ADS-B SLoWC for Head-On MOPS Requirement-Derived Test Vectors

Results for Converging test vectors using ADS-B (Figure 13) are almost identical to the Head-On results. The DAA system was able to keep 99% of the encounters well clear using Sensed ADS-B guidance and 100% of encounters well clear using Mitigated guidance.

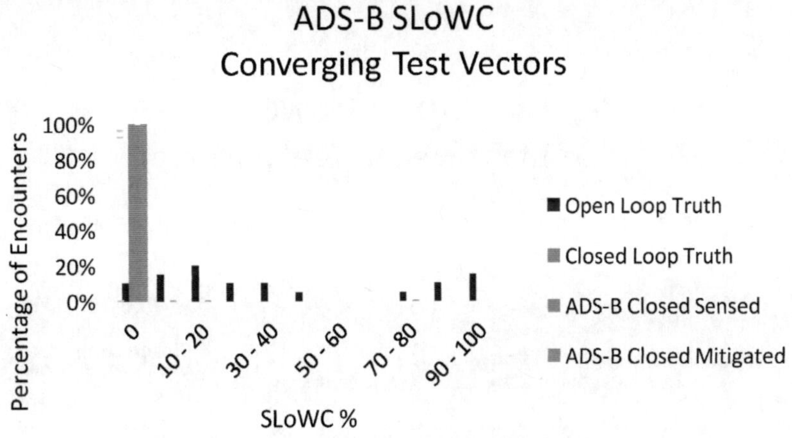

Figure 13. ADS-B SLoWC for Converging MOPS Requirement-Derived Test Vectors

Figure 14 shows results for High Speed test vectors. Closed loop results show 100% of encounters remained well clear when any of the three guidance sources were used (Truth, Sensed, and Mitigated).

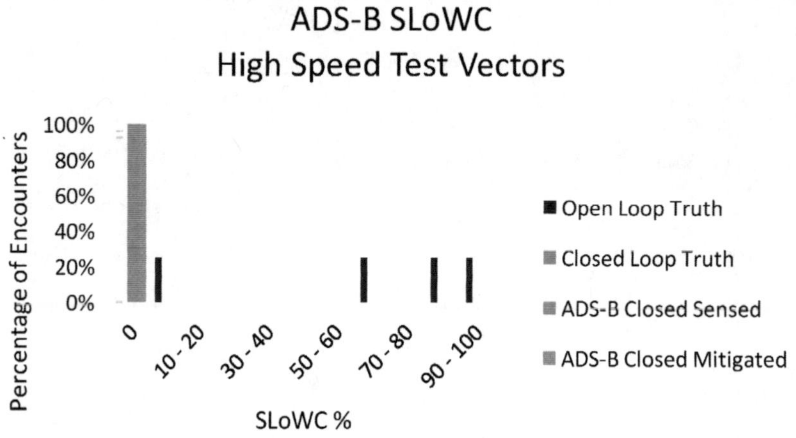

Figure 14. ADS-B SLoWC for High Speed MOPS Requirement-Derived Test Vectors

Figure 15 shows results for Maneuvering test vectors. As with AST and radar, the Maneuvering encounters for MOPS requirements testing proved very difficult, even for ADS-B. Only 13% of encounters had 0% SLoWC using Truth guidance and Sensed guidance results show only 12% of encounters maintained well clear; however, the Mitigated (SUM) approach more than doubled that (32% remained well clear). Overall, the DAA system performed least favorably in the Maneuvering encounters in comparison to the other MOPS requirements-derived encounter geometries.

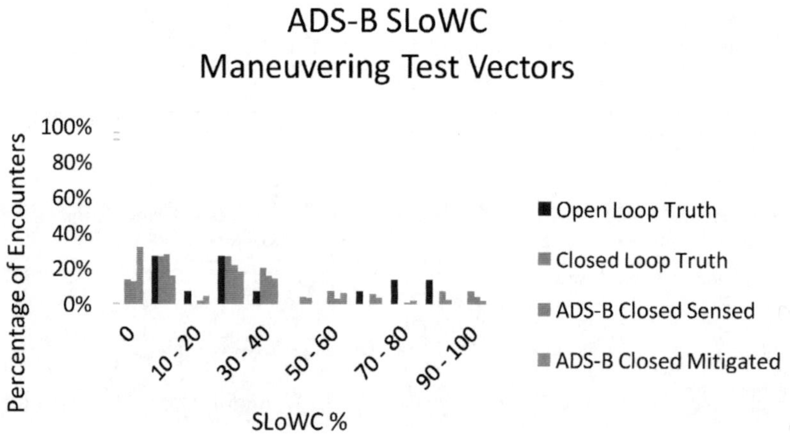

Figure 15. ADS-B SLoWC for Maneuvering MOPS Requirement-Derived Test Vectors

Figure 16 shows results for Overtake test vectors. The DAA system was able to keep 97% of encounters well clear using Sensed guidance and 100% of encounters well clear using Mitigated guidance.

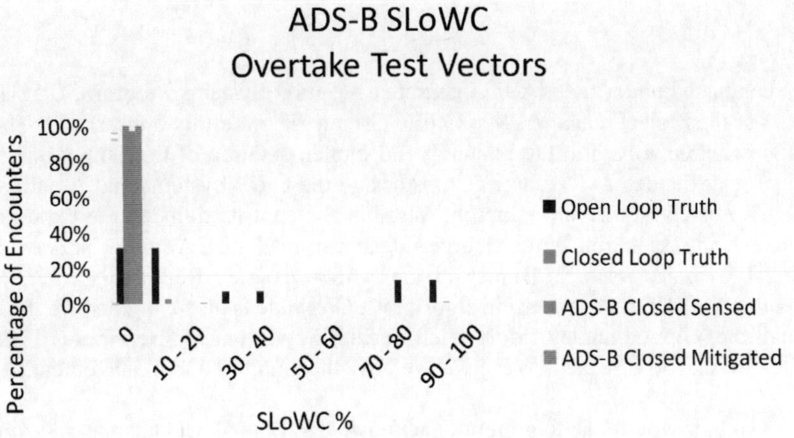

Figure 16. ADS-B SLoWC for Overtake MOPS Requirement-Derived Test Vectors

V. Discussion

The demand for unmanned aircraft in mainstream aviation operations continues to grow. Understanding key detect and avoid system performance capabilities and limitations are essential to developing rules and regulations that allow routine UAS operations but maintain the safety of the National Airspace System. To understand these capabilities and limitations, as part of on-going RTCA SC-228 efforts, NASA Langley Research Center evaluated the Phase I DAA MOPS requirements with end-to-end functionality over a specific set of encounters, in a variety of geometries, and with specific surveillance sensor performance, in order to verify and validate that a MOPS-representative DAA system performs acceptably.

Evaluation results showed that overall, a MOPS-representative DAA system performed within acceptable ranges with few limitations. Values greater than 50% for severity of loss of well clear (SLoWC) occurred in less than 1.5% of total encounters. In particular, the AST sensor produced very inaccurate and noisy intruder tracks that caused multiple issues with SLoWC, and experienced data dropouts in about 70% of the MOPS requirements-derived test vector runs. Results suggest that slow moving aircraft should not depend solely on the AST sensor for lateral maneuvers. It should be noted that all three surveillance sensors (ADS-B, Phase I air-to-air radar, and AST) were modeled to produce data at the minimum specified quality, and can be expected to perform better in the field on average. As expected, the DAA system performed better with ADS-B than with radar or AST. Two other factors affecting DAA system behavior were observed in the pilot model. In some cases, the update delays in the pilot model were longer than a human operator would likely have taken, and the current direction of maneuver was not considered, which led to some cases in which the pilot model tried to steer the UA out of an avoidance maneuver that a human operator would have simply maintained or increased. Also, in some encounters with high uncertainty about the intruder's position and velocity, the sensor uncertainty mitigation (SUM) approach used in this study could cause erroneous guidance and alerting. Taken all together, none of the results of this study revealed surprising or serious problems with a Phase I DAA MOPS-compliant system. Details of the full simulation study can be found in Ghatas, et al. (2017).

Acknowledgments

This work was conducted under the NASA Integrated Aviation Systems Program, UAS Integration in the NAS Project. The support of the Project manager, Ms. Laurie Grindle, is gratefully appreciated. The unfailing and tireless support of the many people involved in the planning and implementation of this simulation study is also gratefully appreciated. Gratitude is extended to Dr. James Comstock of the Crew Systems and Aviation Operations Branch at the NASA Langley Research Center in Hampton, Virginia for contributions to the experiment methodology and conduct. Thanks are extended to Aaron Dutle, George Hagen, Cesar Muñoz, Anthony Narkawicz, and Jason Upchurch of the Safety Critical Avionics Systems Branch at the NASA Langley Research Center for contributions to the development of the DAIDALUS self-separation algorithms. Gratitude is also extended for the tireless efforts of those participating through the NASA Langley Information Technology Enhanced Services (LITES) contract, including: Joel Ilboudo, Kristen Mark, and Elizabeth Weech as well as other support and management personnel.

Additionally, the authors would like to gratefully acknowledge Deepak Chauhan and Kevin Szagala of the FAA's William J. Hughes Technical Center and Dr. Siva Sivananthan, Dr. Akhilesh Shrestha, and Dr. Petro Khomchuk of ARCON Corporation for their help in implementing the sensor and tracker models, Dr. Randal Guendel of MIT Lincoln Lab for his help in implementing the pilot model, Don Walker of the FAA for his leadership, and others in RTCA SC-228 who helped refine the study.

References

Ghatas, R. W., Jack, D. P., Tsakpinis, D., Vincent, M. J., Sturdy, J. L., Muñoz, C., Hoffler, K. D., Dutle, A., Myers, R., DeHaven, A. M., Lewis, T., Arthur, K. E. (2017). Unmanned Aircraft Systems Minimum Operational Performance Standards End-to-End Verification and Validation (E2-V2) Simulation. NASA Technical Memorandum 2017, Hampton, VA.

Jack, D., Hoffler, K., Sturdy, J. (2017) Mitigating the Impact of Sensor Uncertainty on Unmanned Aircraft Operations. NASA CR 2017, Hampton, VA.

RTCA, Inc. (2016). DRAFT Detect and Avoid (DAA) Minimum Operational Performance Standards (MOPS) for Verification and Validation, Version 3.6. RTCA Paper No. 261-15/PMC-1400. RTCA, Inc., Washington, DC. In Press.

The Role of Formal Methods and Continuous Monitoring in the Verification of Complex Air Transportation Systems

Wilson N. Felder[1]
Stevens Institute of Technology, Hoboken, NJ 07030

For a certain class of aerospace systems, such as the U.S. National Airspace System, it is theoretically impossible to fully verify the performance of the system in accordance with its requirements. This can be demonstrated as a consequence of basic system theory, and established in individual cases through the application of formal methods of verification. These systems are termed "complex," and require an additional element of life cycle performance monitoring to protect against unpredictable but inevitable emergent system behaviors. A protocol for the application of continuous monitoring is proposed as an amplification of a previously described framework for the design, development, test and operation of complex aerospace systems.

I. Introduction

This paper reports on the clarification and extension of the use of formal methods and continuous monitoring as part of a general framework for design, development, test, and operation of complex aerospace systems, as described earlier (Felder, 2016b). For some time, I have been advocating this approach, most recently in a special issue of *The ITEA Journal* (Felder, 2015). During the 2016 AIAA Aviation Forum I proposed a specific framework for the Verification and Validation (V&V) of air transportation systems, using the US National Airspace System (NAS) as a notional example. This paper extends and defines two elements of that framework, building on recent advances from the field of commercial software testing, and new findings from the application of formal methods to verification of digital avionics.

As a review, the framework proposed was composed of five key elements:

(1) Life cycle governance;
(2) Model based engineering;
(3) Iterative development;
(4) Continuous monitoring; and
(5) Embedded test tools and formal methods

The impetus for the development of this framework has been the ongoing conversation fostered by the AIAA Complex Aerospace System Exchange (CASE), which seeks to discover best practices for the design, development, test and operation of complex aerospace systems (of which the U.S. NAS is a notable example.)

Work on the elements of the framework continues, with contributions from numerous workers. Both model based engineering and the use of concurrent engineering as a tool for life cycle management were topics of discussion at CASE 2016, and are subjects for chapters in an upcoming book reviewing progress in complex aerospace system design, development, test and operation as reflected through the CASE meetings.

[1] Distinguished Service Professor, School of Systems and Enterprises, Member ATCA, AIAA Fellow

Indeed, the first three elements of the framework are increasingly seen as essential. As a review, it will be helpful to summarize the impetus for, and main aspects of, these three elements.

In this context, "life cycle governance" refers to the need for a continuous, integrated process that guides the evolution of a complex system from research right on through to disposal. This concept addresses our understanding that the speed of advancing technology, highly agile threats, and the continuous adaptive nature of requirements make the traditional stage gate process ineffective and insufficiently responsive (Szajnfarber & Weigel, 2012).

We understand "model based engineering" in a broad and historic context – not the narrowly constrained view that calls for use of a particular system level modeling tool such as SysML. Specifically, my preferred working definition is of a family of models of various types (real and fast time, human in the loop, virtual live and constructive) at various levels of fidelity, that serve as a channel of communication from later phases of earlier system iterations to earlier phases of later iterations.

Finally, the element of "iterative development" is to be understood as the application of the fundamental aspects of agile software development to the slower and lengthier cycles of the

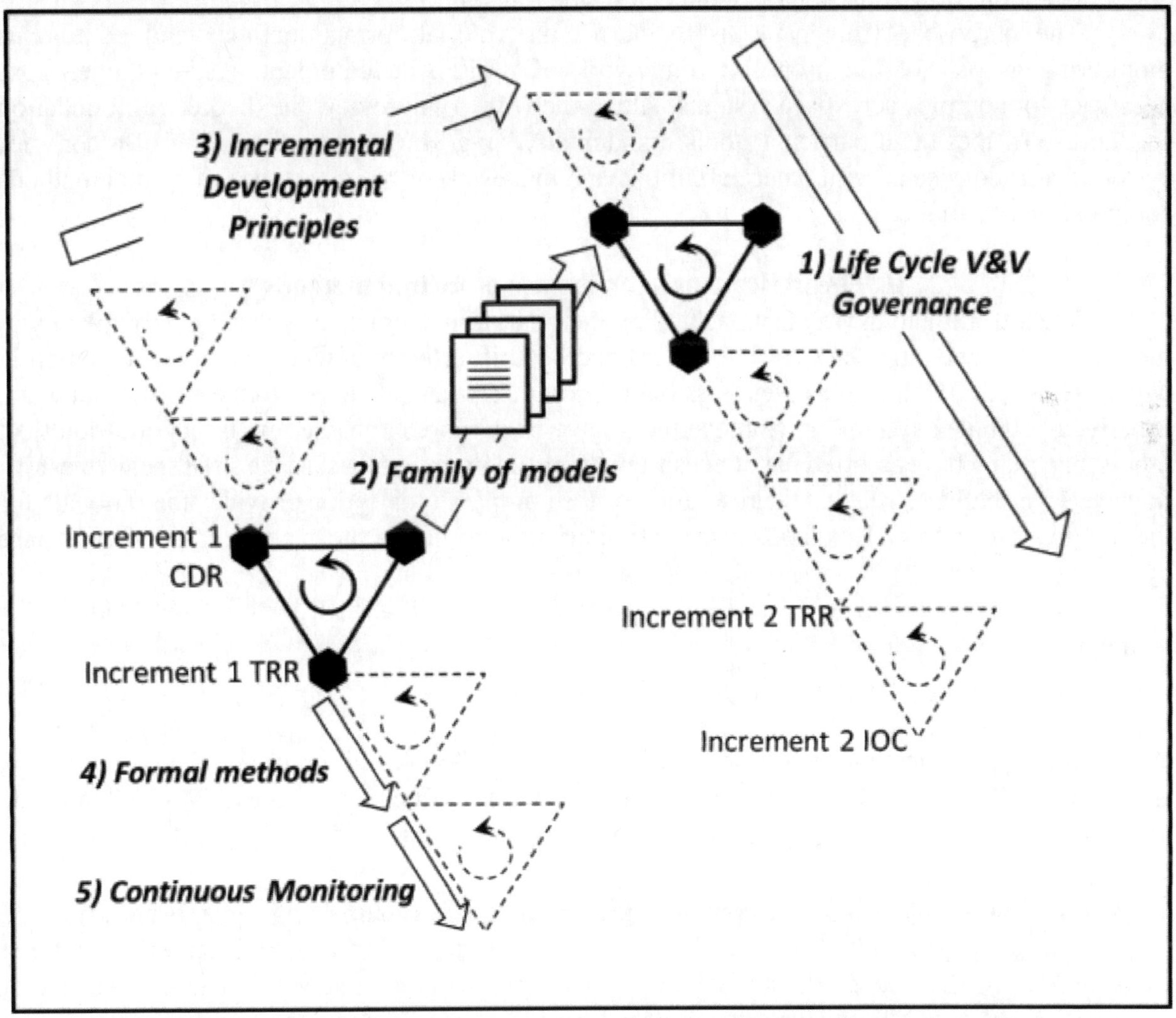

Figure 1: Proposed Framework for Design, Development, Test and Operation of Complex Aerospace Systems (after Felder (2016b) – as Amended)

traditional aerospace industry, which has been described as the incremental commitment spiral model (Boehm, Lane, Koolmanojwong, & Turner, 2014). This concept has been widely applied, both formally and by happenstance, but is recognized as a best practice in the high-speed world of complex system development.

The last two elements of the proposed framework have so far been less well defined, even though they were identified as key complex system verification methodologies at least as early as 2008 (Selberg & Austin, 2008), and in my previous description of the framework (Felder, 2016), were presented as more of a selection of "other" techniques rather than as distinct features. One objective of this paper is to further clarify and extend these ideas as part of the framework structure. Although both approaches are possible aspects of a verification and validation strategy, they differ significantly in their application and timing.

Figure 1 is a revision of the framework description from Felder (2016b) that clarifies the roles and positioning within the life cycle of formal methods and continuous monitoring as elements in the framework. In this view, embedded test tools become not an independent element, but just one aspect of a wide variety of techniques that together can be used to carry out the continuous monitoring function (as will be described in Section IV).

The purpose of this paper is to clarify the role of formal methods and continuous monitoring as part of the proposed framework of design, development, test and operations heuristics for complex aerospace systems. Along the way, I will review the theoretical foundation for the use of these two particular tools, incidentally suggesting a new definition of a complex system that is consistent with other definitions and makes clear the importance of formal methods and continuous monitoring.

II. Theoretical basis for the use of formal methods

We are still a long way from achieving the goal of mathematically defining complexity in the sense envisioned in John Casti's eloquent and beautiful (the only word for it) call to action in *Nature* (Casti, 2001). It is, however, possible to provide a practical answer to the question of what, exactly, a "complex system" is. Some authors have been more concerned with this question than others, and many have asserted that it doesn't matter (and in some sense, as we shall see, it doesn't), however, we shall also see that the resolution of the complexity question provides the rationale for the application of formal methods and continuous monitoring to the management of aerospace systems.

To see how this is so, consider a system of interest S defined in the usual manner, for example by:

$$S: P(\vec{I}) = \vec{O} \supset G\{o_m, o_{m+1}, \dots, o_n\} \qquad (1)$$

Which is just a restatement of von Bertalanffy's original 1950 definition of an open system (Von Bertalanffy, 1950), and which simply says that a system is defined by a process P that operates on a set of inputs $\vec{I}$ to produce a set of outputs $\vec{O}$ of which the subset $\vec{G}$ represents the goals of the system.

We now attempt to create a complete description of the system using some formal language such as the one proposed by Calisi *et al* and which is the subject of further refinement by Andrea Censi at MIT (Calisi, Censi, Iocchi, & Nardi, 2008). We will either succeed or not. If we cannot succeed for whatever reason – architectural or dynamic – then we know that we do not have a full and complete description of the system and in particular that we cannot *a priori* describe either its possible system states, or its behavior when presented with certain inputs (or time series of inputs).

Such a system is not amenable to complete verification prior to deployment, because we cannot know the full extent of its behaviors. We will use the term "complex" to describe such a system. On the other hand, suppose we succeed in building a description of the system using formal language. Such a description is now used to characterize the full response of the system to the complete spectrum of possible inputs (and time phased input strings…), usually by exhaustively tracing individual input values through the model formed by the formal language description of the system. If we are successful, then the design of the system will be "proved," that is, we will have complete understanding of the system's response to each and every possible input value (or time phased string of such values…). If we cannot complete this analysis (either because it's too hard, or for other technical reasons) then we will once again term the system "complex" because its future behavior cannot be predicted.

To put it another way, we will define a system as complex IFF:

1) It cannot be fully described using the syntax of a formal language; or
2) It can be so described but the resulting description cannot be solved.

This definition seems to ring true in the case of almost every example of complex systems that have been suggested in the past. It also admits of the existence of large, convoluted systems such as the Mars Science Laboratory Entry Descent and Landing system (MSL EDL) (Steltzner et al., 2010) and the National Ignition Facility (NIF) (Felder, 2016a), that probably *can* be described using formal language descriptions that are solvable in the formal methods sense. It also might explain why other systems such as the F-35 Joint Strike Fighter, which on the surface might appear simpler and more straightforward than either the MSL EDL or NIF, seem to present difficulties of emergence in their verification campaigns. (That is, they produce results that were not predicted by the design of the system.)

From this discussion, we can now see that any system capable of description using the syntax of a formal language, for which the resulting model can be solved, is amenable to being adequately verified using either formal methods or more traditional test methods during Development Test (DT) and Operational Test (OT). If it cannot be, on the other hand, then it is subject to unpredictable behavior during operation, and other methods must be employed to prevent damaging results from that (unpredictable but inevitable) behavior. In this way, formal methods and continuous monitoring can work together to provide a warning and a safety net for complex systems throughout the life cycle. Note that both these methods are fundamentally all about verification (that is, answering the question "did we build the system right?" as opposed to "did we build the right system?".)

Complex aerospace systems programs that do not follow this protocol: that is, they *neither* are proven using formal methods of verification at the time of Development and Operational Test (DT/OT); *nor* do they put in place a continuous monitoring regime, will inevitably encounter difficulties during operation. As has been previously noted (Felder & Collopy, 2012), the probability of such problems remains high throughout the life cycle because the nature of emergent behavior in complex systems follows power law distributions with fat tails, not the better-behaved Gaussian form, as explained for general audiences by Taleb in his excellent popular account (Taleb, 2007).

As a practical matter, of course, it hardly matters whether the formal methods test is actually applied (although as our facility with these techniques, and the availability of suitable tools, increases, it's probably a good idea to take this step): for the vast majority of digitally rich, distributed systems-of-systems with human interaction and high volume, stochastic, asynchronous digital communication among a changing membership of constituent systems, there will be no easily solvable formal language description possible. As a result, continuous system monitoring is an essential requirement in almost every case. Always understanding, of course, that there are some quite large and convoluted systems (such as MSL EDL and NIF as previously mentioned) that buck this trend.

III. Formal methods in verification of complex aerospace systems

The fundamental concept in formal methods is the idea that a system can be fully described by a "clean algorithm" using some formal language. The formal description can then be proven through formal logic, the conclusion being that a system built according to the formal description is proven to work as designed in all possible cases. The use of formal methods in V&V has been widely reported (Easterbrook & Callahan, 1998). Cofer (2015) has provided an excellent review of the fundamental concepts of formal methods verification and its application to aviation systems. Kuhn *et al* (2002) provide a concise and comprehensive review of the various methods available for this purpose. Such algorithms are relatively straightforward to develop for simple systems. Applying formal methods to more complicated arrays of systems poses more difficulty. Researchers at Rockwell Collins have had significant success in developing formal proof methods for a wide variety of communications and avionics systems and are actively pursuing the application of formal proof methods to larger, distributed and networked examples. When we go beyond a simple, closely contained piece of digital avionics to a wider, networked environment, however, the problem becomes much harder.

Ultimately, the problem with the strict application of formal methods to real systems comes from the nature of the formal proof methodology itself. As described by Gödel's incompleteness theorem, any "sufficiently powerful" formal language is capable of asserting statements that cannot be proven to be true (or false) within the context of the language. In the present case, this means that we cannot be sure that a "clean algorithm" that is provable using formal proof methods can be drawn up for any given system. The importance of Gödel's theorem is not that it is a practical impediment to the application of formal methods of verification, but that because of it, we know that no sufficiently powerful formal language used to describe a given system is guaranteed to create a system description free of unprovable statements. Therefore, the second step, of proving the algorithm through analysis, is required. For a rigorous proof of Gödel's theorem, consult any standard textbook on metamathematics, for example, Kleene (1952). Hofstadter provides a much more readable and entertaining, although still sound, accounting in his *tour-de-force* (Hofstadter, 1980).

Researchers at the Software Engineering Institute (and elsewhere) have had success in applying stochastic approaches in conjunction with formal methods to allow the application of formal methods in less structured environments (Youssef, 2011). It is important to recognize, however, that in these cases, the "proof" aspect of formal methods is necessarily compromised. The stochastic approach essentially works by "pruning" the tree of potential system states, the resulting assertion being some variant of the statement "we cannot prove that the algorithm works in all cases, but we're pretty sure that it does, and we can in fact prove it in an important subset of the problem space." The use of stochastic extensions to formal methods is extremely powerful. It permits evaluations of specific aspects of system performance, such as cybersecurity, and gives good assurance that a particular system has a high probability of surviving a wide range of potentially damaging emergent events.

It should be noted that in Rockwell Collins' lengthy exploration of formal methods, "we have not yet run into Gödel as a limitation...." (Whiting, 2016) However, it is also true that Gödel exists as a potential final barrier to the definitive, closed form determination of system verification. It is for this reason at least that an additional safety net, in the form of continuous system monitoring, is required. It is encouraging that formal methods, perhaps with the addition of stochastic pruning as described above, can at least provide a robust first pass of system verification.

An exciting opportunity for further study of this problem is found in Andrea Censi's work at MIT (Calisi et al., 2008). He has developed a system design tool that can be used to describe, in plain language, the system level requirements for a system, which are then converted into formal language. Software manages the process of description in such a way as to ensure completeness. The requirements are dynamically integrated into a top-level system design. The language is a formal one according to the definitions of formal logic, and is "sufficiently powerful" in the Gödel sense, so it provides an effective platform to determine theoretically whether any given system can, or cannot, be described in a way that makes it appropriate for the application of formal methods absent any additional pruning of system states. A potential experiment would involve coding a candidate system design using Censi's language. One of two things is then true: either (a) such a mapping is feasible, or (b) it is not. If (a), then it is by definition provable by the application of formal methods. If (b) it is not, and therefore will exhibit system states that have not been predicted and may result in damage to the system or unacceptable behavior on the system's part. (actually: if (a), then it is subject to formal proof, which may result in a situation where a given outcome may be provable or not, in which case, see (b)).

Even though in general formal methods are best suited for pre-deployment use, they can also be used to establish benchmarks for system observation during operations. An excellent example of the innovative use of this concept using the modeling tool AADL (developed by the Software Engineering Institute) is described by Hecht (2015).

In the end, however, there are at least three reasons that post-deployment monitoring is required, the first of which is the formal logic limitation described above. A second important factor is the reality that most systems of any interest to the aerospace community are embedded in complex socio-technical systems in which the presence of human operators forces complexity with the attendant presence of emergent behaviors, behaviors that regardless of their genesis, affect the operation of the engineered system and require response by that system. Finally, complex software rich systems exist in surroundings characterized by rapid environmental change, evolving challenges and threats, and emerging use cases. Figure 2 illustrates the application of formal methods and continuous monitoring to provide comprehensive verification over the life cycle.

Significant obstacles remain to the widespread implementation of formal methods, as pointed out by Hall *et al* (Hall, Driscoll, & Schweiker, 2012). Honour (Honour, 2013) too, has pointed out some of the difficulties in system verification that result from distributed systems in a system-of-systems, which is a commonly encountered scenario.

IV. Application of Continuous Monitoring

Formal methods are focused on the design of the system, and are intended to serve as proof that a particular system design will satisfy the system requirements in all possible cases. Continuous monitoring, however, which is also sometimes referred to as recurring surveillance, recognizes that in the case of certain systems of sufficient complexity, formal Development (DT) and Operational Test (OT) campaigns will be unable to fully explore the system's performance envelope, thereby allowing for the possibility of unpredicted (and perhaps unpredictable) failures during the system's operational life. This result is a certainty in the case where the evolving threat environment, spontaneous changes in system use, and other factors, result in unanticipated system use (although strictly speaking, this can only be true if a given system fails to be provable through the application of formal methods, because if the formal methods proof is achievable, then no off nominal behavior is possible). It seems likely that most aerospace systems deployed prior to around the turn of the 21st Century would have been amenable to verification using formal methods, and that only since the widespread interchange of digital data among systems has there been an increase in the prevalence of systems that would be considered "complex" under the metric proposed in Section II. This is, of course, mere conjecture, as is the idea that the source of the complexity resides in the high volume, stochastic, asynchronous nature, and *ad hoc* (i.e. unscheduled) use of such digital interchange. It should be noted that it would be very likely possible to have a system made up of constituent parts (systems or subsystems) each of which could be proven using formal methods of verification, but whose interconnected whole could not be so characterized.

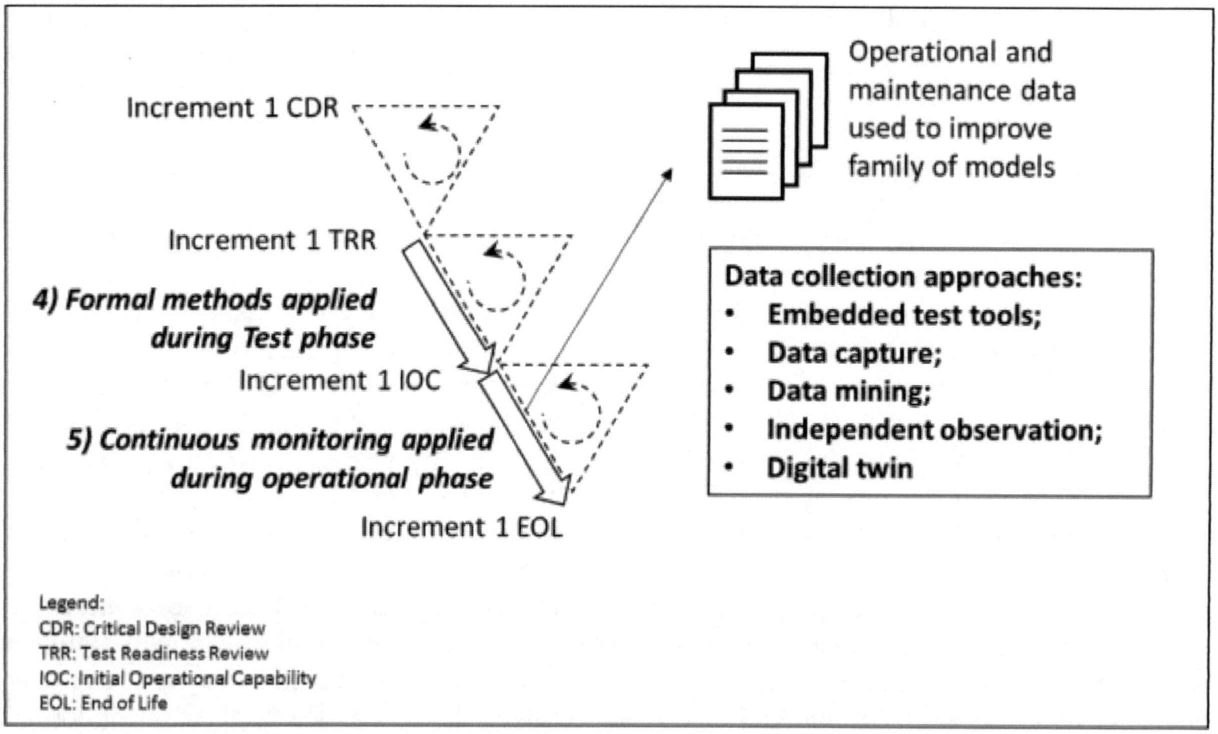

Figure 2: Detail of Framework showing use of Formal Methods and Continuous Monitoring to provide verification over the life cycle

In Figure 2 above a portion of the overall framework shown in Figure 1 is reproduced to clearly show the application of formal methods and continuous monitoring to the test and operation phases of the life cycle.

Formal methods are applied during Development Test (DT – between the Test Readiness Review and Initial Operating Capability milestones), with the objective of (1) providing as complete a "proof" based verification of the system as possible, and (2) establishing whether or not the system can be satisfactorily verified during the formal test phase, or whether continuous monitoring is required during the operational phase in order to protect from emergent effects.

Continuous monitoring is applied, on the other hand, during the operational phase (between Initial Operational Capability and End of Life) and has two major goals: first, to protect against unpredicted but inevitable emergent effects of a damaging nature, and second, to provide an ongoing stream of information to the family of models at the core of the framework in order to permit ongoing improvements to the fielded system (denoted by the circular arrow at the center of the triangle) and to inform the design of the next increment in the system development. We have seen that in almost every case, we will be unable to fully verify the design of a system using formal methods, leading to the need for continuous monitoring of systems performance. In the context of the life cycle, it is best to think of continuous monitoring as a means to protect against that infinitesimally small exposure to unpredicted, and unwanted system states that exist, in a theoretical sense, as a result of the "pruning" process involved with the practical application of formal methods to system verification. Both maintenance and operational data can be used to improve the fidelity of the model(s) and the quality of the resulting designs.

Once we have concluded that continuous monitoring is necessary (and it will always be a useful component of the overall framework because it leads to improvements in the family of models used to characterize future system increments) the process itself can take several different forms, which can include:

- Embedded test tools;
- Monitoring of performance data;
- Analysis of data;
- Independent observation of system performance;
- Maintenance of a "digital twin".

A variety of continuous monitoring tools and techniques are currently used widely across industry and in the aerospace environment generally. All these are useful in the context of a complex system framework, with varying applicability depending on the nature of the system of interest. Figure 2 shows in detail how the various continuous monitoring techniques are employed in practice over the life cycle. A brief review of each of these approaches follows. These techniques have been used fairly widely in software maintenance, under the heading of "recurrent surveillance," as described generally in April *et a l*(2005) and are also reflected in the U.S. Air Force concept of the digital thread/digital twin initiative. I have previously described some of these in connection with the FAA's test programs (Felder, 2016b). These are summarized below for completeness.

At the most fundamental level, embedded test tools have been used widely in the commercial software environment. Under this approach, data is captured during software execution, and used to trigger various responses as required to correct emergent issues. The use of these approaches in aerospace has been less prevalent. Initial results under an ongoing NASA research initiative have proven promising. The use of embedded test is exemplified in the NASA cFS program as reported by McComas *et al* (2015).

A slightly higher level and much more widespread practice involves the collection of performance data on operational systems. This approach is very commonly used by airline maintenance organizations to monitor system performance aboard aircraft in near real time in order to schedule preventative and corrective maintenance. The data are downloaded to airline technical operations via the ACARS digital data link system.

The FAA has used continuous monitoring to ensure the satisfactory performance of several systems acquired as services. Because the verification of these systems through testing is proprietary, and therefore beyond the agency's visibility, performance data are routinely collected on the Wide Area Augmentation System (WAAS) (McHugh, 2005), the Automated Dependent Surveillance System – Broadcast (ADS-B) system, and the Flight Service Station program. The importance of continuous monitoring in these cases has also been the subject of Department of Transportation Inspector General reporting (Guzzetti et al., 2011).

The use of independent observation to monitor performance has been used with the FAA's Reduced Vertical Separation Minima (RVSM) program (Falk, Gonzalez, & Perez, 2010).

Although a vital part of the U.S. National Airspace System, this system has no ground infrastructure, and no Government owned software or hardware. In this instance, the agency uses a network of independent observation stations in which the altitude of the aircraft, as transmitted from the on-board precision altimeter, is compared to an independently measured altitude derived from primary radar triangulation, or other methods.

In addition to simply recording the data observed about a given system's performance, data mining and analysis can be used to derive secondary results relative to system operation. To a certain extent this is already being done by the airline technical operations systems described earlier. An intriguing application of this continuous monitoring concept has been described by Normann (2015).

In all of these cases, data collected from, measured on, or extracted from operational systems is available for use by the central collection of models that can no be used to develop fixes to ongoing maintenance issues, or to design the next increment in a family of systems. A distinctive use of continuous monitoring to improve the family of models used for design of future system increments is the Air Force Digital Thread concept as described by Kraft (2015) (Kraft, 2015), and by West and Pyster (2015). The Digital Twin concept is also extensively reviewed in the recently published review of findings from the series of CASE exchanges (Grieves & Vickers, 2016). The Air Force actually recognizes two linked concepts – the digital thread and the digital twin – where the "twin" is a powerful unitary model that mimics the physical system with high fidelity, and the "thread" is a framework provided to build, maintain and operate the "twin." The digital twin concept is prevalent in manufacturing where it has not generally been applied to complex systems. The "digital thread" idea is similar in intent to the FAA's family of models – sometime called a System of Systems Assessment Platform (SOSAP)– described in my earlier framework paper (Felder, 2016b).

V. Conclusions

The goal of this paper has been to clarify the roles of formal methods and continuous monitoring as elements in an overall framework for the design, development, test, and operation of complex aerospace systems. Each of these are intended as tools for verification of system performance but they act in different ways. Formal methods are a more theoretically based version of traditional development and operational test methods. If successful, they can be used to prove that a given system behaves as designed for all possible variations of system inputs. On the other hand, if no formal methods based verification process is possible, we are faced with a complex system for which we cannot predict all possible outcomes. In this case, a continuous monitoring regime is required to protect against unpredictable but inevitable emergent effects with potentially detrimental impact to the system.

With the greater definition of the role to be played by formal methods and continuous monitoring in the design, development, test and operation of complex aerospace systems such as the U.S. National Airspace System, we now have a much stronger and more robust proposed framework for the successful and confident deployment of operational systems. In addition, we have a means for determining whether a given system is likely to require continuous monitoring because of inherent complexity. The result, as depicted in Figures 1 and 2, represents "version 2.1" of the proposed framework.

References

April, A., Huffman Hayes, J., Abran, A., & Dumke, R. (2005). Software Maintenance Maturity Model (SMmm): the software maintenance process model. *Journal of Software Maintenance and Evolution: Research and Practice*, *17*(3), 197–223.

Boehm, B., Lane, J. A., Koolmanojwong, S., & Turner, R. (2014). *The incremental commitment spiral model: Principles and practices for successful systems and software*. Addison-Wesley Professional.

Calisi, D., Censi, A., Iocchi, L., & Nardi, D. (2008). OpenRDK: a modular framework for robotic software development. In *2008 IEEE/RSJ International Conference on Intelligent Robots and Systems* (pp. 1872–1877). IEEE.

Casti, J. L. (2001). Formally speaking. *Nature*, *411*(6837), 527–527.

Cofer, D. (2015). Taming the Complexity Beast. *The ITEA Journal*, *36*(4), 313–318.

Easterbrook, S., & Callahan, J. (1998). Formal methods for verification and validation of partial specifications: A case study. *Journal of Systems and Software*, *40*(3), 199–210.

Falk, C., Gonzalez, J., & Perez, J. (2010). Using Automatic Dependent Surveillance-Broadcast Data for Monitoring Aircraft Altimetry System Error. In *Paper no. AIAA 2010-8165* (pp. 1–12). Toronto, Ontario Canada: AIAA.

Felder, W. N. (2015). System Engineering, Reliability, Life Cycle Support and Testing: a New Theory of Test for Complex Systems. *The ITEA Journal*, *36*(4), 259–263.

Felder, W. N. (2016a). Systems engineering best practices on the National Ignition Facility. Presented at the 2016 Conference on Systems Engineering Research, Huntsville, AL.

Felder, W. N. (2016b). The U.S. National Airspace System: a Model for Verification and Validation of Complex, Distributed Systems-of-systems. In *AIAA 2016-3152* (p. p.3152). American Institute of Aeronautics and Astronautics. Retrieved from http://dx.doi.org/10.2514/6.2016-3152

Felder, W. N., & Collopy, P. (2012). The elephant in the mist: What we don't know about the design, development, test and management of complex systems. *Journal of Aerospace Operations*, *1*(4), 317–327.

Grieves, M., & Vickers, J. (2016). Digital twin: Mitigating unpredictable, undesirable emergent behavior in complex systems. In *Transdisciplinary Perspectives on Complex Systems* (pp. 85–113). Springer.

Guzzetti, J. B., DeWeese, B., Ward, C., Bond, G., Thiel, H., Nossaman, A., & Azuolas, A. (2011). *FAA Oversight Is Key for Contractor-Owned Air Traffic Control Systems That Are Not Certified* (No. AV-2011-149,) (p. 21pp.). DOT OIG.

Hall, B., Driscoll, K., & Schweiker, K. (2012). Verification and validation of distributed flight critical systems. In *2012 IEEE/AIAA 31st Digital Avionics Systems Conference (DASC)* (p. 9D3–1). IEEE.

Hecht, M. (2015). Verification of Software Intensive System Reliabiity and Availability Through Testing and Modeling. *The ITEA Journal*, *36*(4), 304–312.

Hofstadter, D. C. (1980). *Godel, Escher and Bach: An Eternal Golden Braid*. New York: Random House.

Honour, E. (2013). Verification and Validation Issues in Systems of Systems. *arXiv Preprint arXiv:1311.3626*.

Kleene, S. C. (1952). *Introduction to metamathematics, D. Van Nostrand Company, New York*. New York: D. Van Nostrand Co. inc.

Kraft, E. M. (2015). HPCMP CREATE TM-AV and the Air Force Digital Thread. In *Proceedings of the 53rd AIAA Aerospace Sciences Meeting, AIAA 2015* (Vol. 42).

Kuhn, D. R., Chandramouli, R., & Butler, R. W. (2002). Cost effective use of formal methods in verification and validation. In *Foundations' 02 Workshop* (pp. 22–23).

McComas, D. C., Strege, S. L., Carpenter, P. B., & Hartman, R. (2015). Automated Test for NASA cFS. *The ITEA Journal*, *36*(4), 278–287.

McHugh, T. (2005, June). *WAAS Program Test Overview and In-Service Monitoring*. FAA WJHTC Media Briefing.

Normann, B. (2015). Continuous System Monitoring as a Test Tool for Complex Systems of Systems. *The ITEA Journal*, *36*(4), 298–303.

Selberg, S., & Austin, M. A. (2008). Toward an evolutionary system of systems architecture. In *Proceedings of Eighteenth Annual International Symposium of The International Council on Systems Engineering (INCOSE 2008), Utrecht, The Netherlands*. Citeseer.

Steltzner, A. D., Burkhart, P. D., Chen, A., Comeaux, K. A., Guernsey, C. S., Kipp, D. M., … others. (2010). Mars science laboratory entry, descent, and landing system overview.

Szajnfarber, Z., & Weigel, A. L. (2012). Managing complex technology innovation: the need to move beyond stages and gates. *International Journal of Space Technology Management and Innovation (IJSTMI)*, *2*(1), 30–48.

Taleb, N. N. (2007). The Black Swan: The Impact of the Highly Improbable.

Von Bertalanffy, L. (1950). The theory of open systems in physics and biology. *Science*, *111*(2872), 23–29.

West, T. D., & Pyster, A. (2015). Untangling the Digital Thread: The Challenge and Promise of Model--- Based Engineering in Defense Acquisition. *INSIGHT*, *18*(2), 45–55.

Whiting, M. (2016, October). Personal Communication. A forthcoming publication will be cited in the final version of this paper.

Youssef, H. (2011). *Verification and Validation of Complex Systems*. SAE Technical Paper.

UNMANNED AIRCRAFT SYSTEMS: THE NAS AT A CROSSROADS

Tom Farrier
Senior Safety Analyst
JMA Solutions, LLC
April 6, 2017

Abstract

The rapid growth of the unmanned aviation sector has game-changing implications for every stakeholder associated with the operation and future of the National Airspace System (NAS). The sector's economic clout is growing, and the existing body of rules and accommodations that enable unmanned aircraft systems access to the NAS is struggling to keep up with demand, innovations, and the ever-widening range of business opportunities to which unmanned aircraft are being applied. At the same time, hazards created by unmanned aircraft operated outside the existing regulatory structure threaten the safety of manned aircraft and the reputation and opportunities of commercial UAS operators. Re-thinking of the entire notion of unmanned aircraft system "integration" in the NAS is needed to consolidate and build upon the initial progress made in advancing the evolution of the sector as a whole.

Introduction

Unmanned aircraft systems (UAS) are often thought of as new factors to be dealt with simply in terms of their actual or potential impact on the safety, capacity, or efficiency of the existing National Airspace System (NAS). However, it is increasingly important for UAS to be acknowledged and accommodated as a user community with the potential to make significant economic contributions to the nation in its own right.

Many challenges remain for regulators to meet UAS operator demand for ever-broader access to the NAS, side by side with current manned aircraft users. This raises the question: *At what point will the NAS itself need to change to accommodate the needs of the unmanned aviation sector, and what form might UAS-oriented changes take?*

Unmanned Aircraft Systems as "Game-Changers"

In June of 2016, the formal publication of Part 107 of Title 14 of the Code of Federal Regulations (14 CFR) cleared the way for tens of thousands of newly certificated "small unmanned aircraft system (sUAS) operators to make use of unmanned aircraft weighing less than 55 pounds for all manner of commercial uses." Already booming in terms of both registrations and requests for authorization to operate in controlled airspace, the small UAS sector has the potential to grow far beyond its current level of activity.

The increasing sophistication of larger unmanned aircraft, emerging technologies aimed at enabling UAS operations far beyond the visual line of sight of their pilots in command, and the many commercial uses that unmanned aircraft with greater payload capacity could be applied against, all suggest that we are only seeing the beginning of a true revolution in how the NAS is used and how it will need to be shared in the coming years.

Even as these trends gain momentum, many current users of the NAS have resisted the growing presence of UAS. While their primary and legitimate concern is for the safety of other aircraft, there are also some commercial activities long supported by manned aircraft that are likely to become increasingly attractive for unmanned platforms as well.

For these reasons, unmanned aircraft systems may be seen as an ideal example of what is known as "disruptive technology." [1] This term has been variously defined with respect to its relationship with different industries. However, one commonly used version in the context of unmanned aviation – paraphrased – is, "A technology that displaces or has the potential to displace an established technology, or that creates a completely new industry."

Writing in the *Harvard Business Review* in December of 2015, Clayton Christensen described how disruption affects the status quo:

> "Disruption" describes a process whereby a smaller company with fewer resources is able to successfully challenge established incumbent businesses. Specifically, as incumbents focus on improving their products and services for their most demanding

[1] Clayton M. Christensen, *The Innovator's Dilemma: When New Technologies Cause Great Firms to Fail* (Boston, MA: Harvard Business Review Press, 1997), p. xix.

(and usually most profitable) customers, they exceed the needs of some segments and ignore the needs of others. Entrants that prove disruptive begin by successfully targeting those overlooked segments, gaining a foothold by delivering more-suitable functionality—frequently at a lower price.[2]

With the above precepts in mind, it becomes necessary to take a step back from how the current NAS works and whom it serves, and instead consider what the rise of unmanned aviation might suggest about its future direction and priorities.

The Shifting Economic Landscape

For more than forty years, the FAA has made regulatory and investment decisions based, in part, on economic analyses. These revolve around three program priorities:

- The need to maintain and update air traffic communication, navigation, surveillance and management services;
- The development and enforcement of regulations for the safe and efficient operation of the NAS; and
- The administration of the Airport Improvement Program (AIP).

Analyses aimed at servicing these program areas are based on five "benefit" domains: safety, capacity increases (to minimize delays), prevention of flight disruptions (especially due to adverse weather), overall cost savings (including "cost avoidance"), and a catch-all "other" category.

How does the unmanned aviation sector fit into this model, as currently constituted? The answer is: not very well. Over time, the safety, capacity, and efficiency of the NAS itself have been measured by the extent to which the needs of the "aviation industry" – particularly the commercial air carriers – are met. This has been a reasonable approach due to the significant contributions to the annual Gross Domestic Product (GDP) that the industry has made over time. The *Airlines for America* (A4A) estimates that these economic impacts represent roughly 5% of GDP and 10 million jobs over at least the past decade.[3] In addition, the Air Transport Action Group places the industry's worldwide GDP contribution at 3.5%.[4]

Table 6 in the November 2016 edition of the FAA's annual *Economic Impact of Civil Aviation on the U.S. Economy* (on the following page) bears out A4A's claims regarding the importance of the aviation industry's contributions to the economy. What is *not* included in this table – nor is it in any way yet factored into the FAA's overall decision-making calculus – is how *unmanned* aviation is beginning to make its own distinct contributions to the nation's economic health. However, some roughly comparable data is contained in a study commissioned by the Association of Unmanned Vehicle Systems International (AUVSI) in 2013, suggesting those contributions could be substantial.

[2] https://hbr.org/2015/12/what-is-disruptive-innovation (accessed March 31, 2017).

[3] A4A Presentation: *Industry Review and Outlook* (March 8, 2017) - http://airlines.org/blog/dataset/a4a-presentation-industry-review-and-outlook/ (last accessed March 31, 2017).

[4] http://aviationbenefits.org/economic-growth/value-to-the-economy/

Table 6. U.S. Civil Aviation: Growth of Total Output, Earnings, and Jobs (Real)

	Output (2012 $Billions)			Earnings (2012 $Billions)			Jobs (Thousands)		
Description	2012	2014	Percent Change	2012	2014	Percent Change	2012	2014	Percent Change
Airline Operations	300.4	314.3	4.6	73.4	76.8	4.6	1,412	1,528	8.2
Airport Operations	70.9	73.4	3.6	22.6	23.4	3.6	490	524	7.1
Civilian Aircraft Manufacturing	122.7	138.9	13.2	30.7	34.8	13.2	547	640	17.1
Civilian Aircraft Engine and Engine Parts Manufacturing	15.9	17.0	6.9	3.8	4.1	6.9	71	78	10.6
Civilian Other Aircraft Parts and Equipment Manufacturing	63.6	71.4	12.3	16.0	17.9	12.3	311	361	16.1
Civilian Avionics Manufacturing	22.9	21.9	-4.3	5.8	5.5	-4.3	112	111	-1.0
Civilian Research and Development	22.7	31.3	38.3	7.2	10.0	38.3	135	193	43.0
Air Couriers	58.1	60.6	4.4	17.6	18.4	4.4	468	505	8.0
Visitor Expenditures	712.4	745.4	4.6	206.7	216.3	4.6	5,649	6,114	8.2
Travel Arrangements	16.4	17.9	9.0	4.8	5.2	9.0	112	126	12.7
Subtotal - Commercial	1,406.0	1,492.3	6.1	388.5	412.3	6.1	9,305	10,161	9.4
General Aviation Operations	34.6	38.5	11.4	8.5	9.4	11.4	163	183	12.2
GA Aircraft Manufacturing	20.0	29.5	47.3	5.0	7.4	47.3	89	132	48.4
GA Visitor Expenditures	11.7	11.6	-0.9	3.4	3.4	-0.9	93	93	-0.2
Subtotal - General Aviation	66.3	79.6	20.1	16.9	20.3	20.0	345	408	18.2
Total Impact	1,472.3	1,571.9	6.8	405.4	432.6	6.7	9,650	10,589	9.7

That report – *The Economic Impact of Unmanned Aircraft Systems Integration in the United States* – makes (and defends) the following conclusions:

1. The economic impact of the integration of UAS into the NAS will total more than $13.6 billion in the first three years of integration and will grow sustainably for the foreseeable future, cumulating to more than $82.1 billion between 2015 and 2025.
2. Integration into the NAS will create more than 34,000 manufacturing jobs and more than 70,000 new jobs in the first three years.
3. By 2025, total job creation is estimated at 103,776.
4. The manufacturing jobs created will be high paying ($40,000) and require technical baccalaureate degrees.
5. Tax revenue to the states will total more than $482 million in the first 11 years following integration (2015-2025). [5]
6. Every year that integration is delayed, the United States loses more than $10 billion in potential economic impact. This translates to a loss of $27.6 million per day that UAS are not integrated into the NAS.

[5] The Economic Impact of Unmanned Aircraft Systems Integration in the United States. Retrieved from http://robohub.org/_uploads/AUVSI_New_Economic_Report_2013_Full.pdf

SAGE Publication's *Business Researcher* rendered the AUVSI report's data as an easier-to-understand graph ("Projected direct spending and economic impact of commercial drone use, in $U.S. billions, 2015–25")[6]:

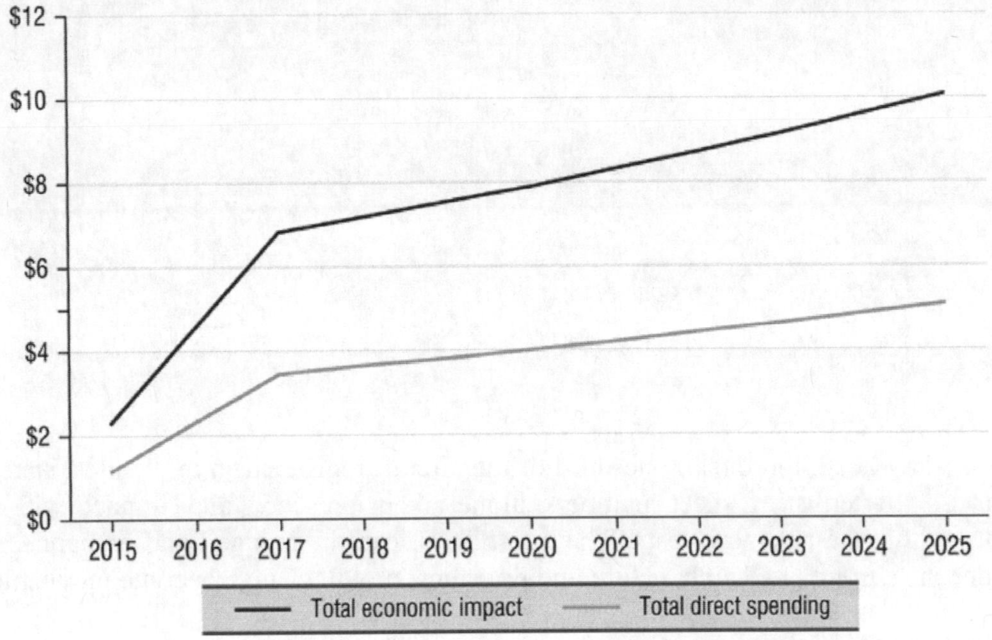

Some of the implications of the above with respect to the FAA's priorities in NAS infrastructure and regulation are potentially significant. Regardless of whether UAS are treated as a separate category or are clustered with "general aviation" operations (whose GDP contributions would grow by more than 50% at a stroke with such a bookkeeping change), unmanned aviation may require substantially more deference in the FAA's future infrastructure and regulatory calculus.

Projected Sales Growth of the Unmanned Sector

No treatment of this subject would be complete without at least a brief look at just how many unmanned aircraft are likely to be plying the skies in just a few years' time. The following graphic[7] was assembled from AUVSI data by *Cognizant*, a business and technology-oriented analytics firm:

[6] "Business Researcher." http://businessresearcher.sagepub.com/sbr-1775-98070-2712950/20160118/economic-impact-of-commercial-drones

[7] https://www.cognizant.com/InsightsWhitepapers/drones-the-insurance-industrys-next-game-changer-codex1019.pdf

Projected Annual Sales of Unmanned Vehicles

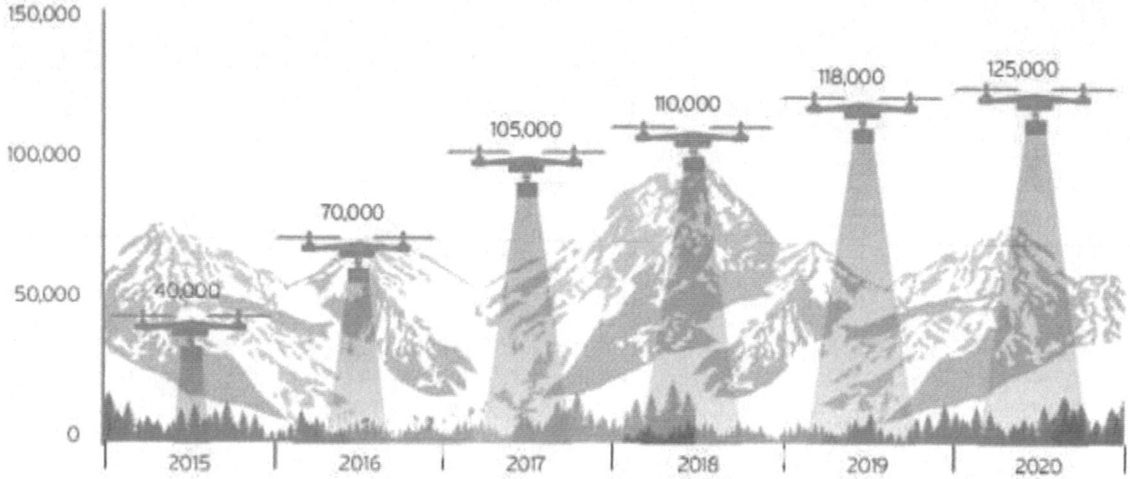

In other words, by 2020, more than one-third of the aircraft registered in the United States could be unmanned.[8] In terms of sheer numbers alone, their presence and impact will expand substantially within the next few years. The question is, can the manned and unmanned sectors coexist under the current NAS architecture and priorities, or will change become inevitable?

Adapting to Disruption

Over the past two years, the growth of unmanned aircraft systems as part of the aviation user community has been nothing short of explosive. Demand for the economies, flexibility, and new uses of UAS has driven the sector's rapid growth. In turn, this growth requires airspace within which UAS can operate without increasing risk to current users of the NAS.

The FAA has traditionally seen the aviation industry – especially the major air carriers – as its "customers" and has sought to meet their needs over time. Today, unmanned aviation is impacting the long-standing relationship between "air commerce" and "air safety," while at the same time being guided by those same two principles. It represents a new and disruptive notion of air commerce itself.

The proponents of unmanned aviation are increasingly challenging the notion that their activities adversely affect existing stakeholders from a safety perspective, suggesting that their impact is minimal and acceptable if they are given due consideration in the allocation of airspace and resources to support the ongoing development of their industry.

As the expansion of the unmanned aviation sector has soared, permanent changes in the operating environment may be needed to accommodate it. This is forcing both the regulators of and the participants in the present-day aviation system to confront a landscape that is shifting beyond what were once considered established and immutable norms. Both will need to adapt to unmanned aircraft technology (including its operational limitations), the economic potential it carries with it,

[8] https://www.rita.dot.gov/bts/sites/rita.dot.gov.bts/files/publications/national_transportation_statistics/html/table_01_11.html

and the technology's impact on "air safety" itself. Indeed, all may need to consider changes in the operating environment in order to accommodate it.

Regulations for sUAS

As mentioned above, Part 107 has enabled small unmanned aircraft system (sUAS) operations to be carried out with minimal and acceptable increase in risk to other aircraft and the general public. It does this by limiting the maximum altitude of such operations to less than 400 feet above ground level (AGL) and the maximum speed of small unmanned aircraft to less than 88 knots. It also limits the operations themselves to day visual meteorological conditions, within the visual line of sight of the pilot in command, and in uncontrolled airspace.

At the same time, these operating rules do not allow commercial sUAS operators to do the kinds of things they want to do for business purposes. In effect, Part 107 is largely a codification of 35-year old guidelines for the operation of model aircraft contained in a 1980s-vintage advisory circular,[9] with added content specific to "remote pilot certification" and a few rules derived from Part 91.

Part 107 was necessary because small unmanned aircraft systems simply cannot comply with significant requirements contained in Part 91 applicable to "aircraft," regardless of them being manned or unmanned, large or small. Even at that, as published, Part 107 contains an opportunity for sUAS operators to obtain relief from virtually all of the operating restrictions contained in the rule through the Certificates of Waiver or Authorization (COA) process – which has become essential to realizing their vision of "commercial operations."

There are three principal reasons why the final rule reads as it does:

1 - External Stakeholders

About six years ago, Congress developed an interest in allowing unmanned aviation greater access to the NAS. By early 2012, this manifested itself in formal legislation embedded in Subtitle B of the *FAA Modernization and Reform Act of 2012* (P.L. 112-95, often referred to as "FMRA"), which contains a series of sections specific to unmanned aircraft systems. The language of these sections explicitly required fast action from the FAA on a variety of fronts, including rulemaking.[10]

2 - UAS Limitations versus Regulatory Objectives

The initial movement toward sUAS rulemaking came with the recommendations of the "Small Unmanned Aircraft System Aviation Rulemaking Committee" (sUASARC), which issued its recommendations in April of 2009.[11] While the sUASARC report devoted most of its content to matters of safety, the first sentence in its Foreword makes the purpose of their work clear: "The Small Unmanned Aircraft System (sUAS) Aviation Rulemaking Committee (ARC) was focused on making recommendations for Federal regulations for the operation of civil (commercial) sUAS."

---

[9] AC 91-57, *Model Aircraft Operating Standards* (originally issued June 9, 1981; revised January 2016 as AC 91-57A).

[10] *See* FMRA Section 332 *et seq. at* https://www.faa.gov/uas/media/Sec_331_336_UAS.pdf

[11] https://www.faa.gov/regulations_policies/rulemaking/committees/documents/media/SUASARC-4102008.pdf.

Part 107's Subpart D ("Waivers), Section 107.200 ("Waiver policy and requirements") states, "The Administrator may issue a certificate of waiver authorizing a deviation from any regulation specified in §107.205 if the Administrator finds that a proposed small UAS operation can safely be conducted under the terms of that certificate of waiver." This includes provisions intended to reduce the likelihood of hazards associated with the most prominent and frequently referred to limitation of unmanned aircraft – the inability of a UAS pilot in command to see and avoid other aircraft and conform to right-of-way requirements – all of which can be eliminated with the stroke of a waiver pen.

3 - Constraints on Needed Regulation

The rulemaking team that developed Part 107 was also constrained by a different aspect of FMRA: a legislative ban on the FAA's ability to regulate an entire family of small unmanned aircraft systems, namely, *model aircraft*.[12]

Model aircraft are listed as a subset of the larger family of "small unmanned aircraft" in the legislation for good reason: a Part 107-operated sUAS is physically indistinguishable from one flown for "hobby and recreational purposes." It is remarkably easy to obtain and operate a small unmanned aircraft – buy it and fly it. The FAA has called for all small unmanned aircraft to be registered regardless of their planned use, but the mechanism for doing so walks a fine line that skirts the "no rule applicable to model aircraft" mandate in FMRA.

At the same time, on a daily basis, the FAA receives "sighting" reports of small unmanned aircraft operating in airspace where they are prohibited. They are almost always far above Part 107's 400-feet AGL ceiling, and frequently are too close to working airfields for comfort (or safety). To date, *none* of these operations have been traceable to a Part 107 operator who was granted permission to operate in the affected airspace.

The *Academy of Model Aeronautics* (AMA), which has sought to be recognized as one of the "community-based organizations" referred to in FMRA, is noticeably silent on the question of operating altitude in its "National Model Aircraft Safety Code."[13] The AMA's safety guidelines only require model aircraft pilots to "Not fly higher than approximately 400 feet above ground level within three (3) miles of an airport without notifying the airport operator."

Under FAA rules, a model aircraft flown in compliance with the parameters established by FMRA is largely free to operate at will, as long as it meets two conditions:

- The aircraft is operated in a manner that does not interfere with and gives way to any manned aircraft; and
- When flown within 5 miles of an airport, the operator of the aircraft provides the airport operator and the airport air traffic control tower (when an air traffic facility is located at the airport) with prior notice of the operation.[14]

---

[12] FMRA Section 336 - h*ttps://www.faa.gov/uas/media/Sec_331_336_UAS.pdf*

[13] http://www.modelaircraft.org/files/105.pdf.

[14] 14 CFR Part 101, Subpart E, "Special Rule for Model Aircraft."

The bottom line is that if it suits an individual's purpose to fly at an altitude higher than prescribed by Part 107, that individual can do so at will by asserting that the operation is for "hobby and recreation." In effect, the FAA has been obliged to treat aircraft with equal capabilities and presenting equal risk to the NAS if operated improperly as somehow being different based on the intent of the operator. One must consider the safety implications of this standard that potentially allows a path to non-compliance.

"See and Avoid" in the NAS

Beyond the extent to which it is addressed by Part 107 (or can be waived under that Part's authority), the "see and avoid" challenge remains unresolved. Most of the alternate means of compliance (AMOC) with 14 CFR §91.113 are likely to be impractical to implement on a long-term, widespread basis. Government/industry efforts to establish firm standards for "detect and avoid" (DAA) systems – both ground-based and airborne – continue to move forward. However, the former tend to be expensive and highly localized in their coverage, and the latter are likely to be impractical for small UAS (being complex, potentially heavy, power-hungry and expensive on a per unit basis).

Both Congress and the FAA have defined unmanned aircraft as "aircraft," meaning their pilots are just as responsible for following the "rules of the road" as their manned aircraft counterparts. Title 49 of the United States Code (49 U.S.C.), Section 40103 states, "The Administrator of the Federal Aviation Administration shall develop plans and policy for the use of the navigable airspace and assign by regulation or order the use of the airspace necessary to ensure the safety of aircraft and the efficient use of airspace."

As the numbers and uses of UAS continue to grow, keeping manned and unmanned aircraft safely separated is likely to become an ever-growing challenge. This fact, coupled with the steadily growing economic benefits resulting from unmanned aircraft applications in a host of different industries and applications, is likely to challenge the current paradigm of manned aviation taking absolute precedence over unmanned aviation throughout the NAS, even in nominally "visual" operations.

At the same time, the *manned* aircraft pilot's obligations under the "see and avoid" requirement are becoming increasingly difficult to carry out in the presence of unmanned aircraft operations thanks to the size and speeds of many of the newest and most popular drones. In many instances, such as is the case in trying to avoid "bird strikes," small unmanned aircraft cannot be seen until it is too late to take effective avoidance action. To date, a viable set of solutions to these challenges has not been found.[15]

This issue has given rise to conversations regarding two considerations related to it: ensuring that UAS pilots in command (PIC) actually participate in the see-and-avoid process, and keeping unmanned aircraft separated from each other.

15

Automated vs. Autonomous Operations

Terminology often tends to be imprecise in discussing distinctions between *automated* versus *autonomous* UAS operations. The former is a natural outgrowth of technology; the latter represents a specific type of flight activity that raises significant questions of safety as the NAS is currently constituted.

Automation supports unmanned aviation in a variety of forms. Even the simplest multi-rotor small unmanned aircraft can be affordably equipped with rotor speed control and altitude monitoring that substantially reduces the workload previously associated with flying any kind of radio-controlled helicopter or fixed-wing aircraft.

On the other hand, the notion of *autonomous* UAS operations across the "size" spectrum of unmanned aircraft has always been suspect in regulatory circles. Conceptually, a purely "autonomous" aircraft is one that is launched, guides itself to a particular location, performs an assigned task (e.g., delivery, aerial photography, aerial application), and returns to its point of origin or a designated alternate location *without the possibility of human intervention in the operation*.

The qualifier at the end of that description is crucial. Years of research into "human in the loop" (HITL) and "human on the loop" (HOTL) UAS concepts has been conducted with the goal of minimizing workload, maximizing efficiency, and reducing the opportunity for human error to adversely affect a given operation. However, both HITL and HOTL provide for the possibility of human intervention into the conduct of a given UAS flight at any time. The key distinction is the fact that autonomous flights are self-executing, while others are not.

Other concepts that rely heavily on automation, such as "swarming" or other multi-aircraft operations managed by fewer pilots than aircraft involved in a given operation, are seen with a degree of wariness. However, these are not "autonomous" thanks to the supervision each aircraft receives and the possibility of immediately terminating any such operation where the safety of the surrounding airspace is in danger of being compromised. Given the inability of most unmanned aircraft to autonomously detect and avoid in-flight conflicts, a valid safety case for allowing autonomous operations may not exist at this writing.

Traffic Monitoring versus Traffic Management

One capability intended to supplement DAA-based AMOCs that has been gaining traction in research and development in recent years is collectively referred to as "unmanned traffic management" (UTM). As with many aspects of unmanned aviation, this is a useful concept possessing a name that resembles a manned aviation functionality, but which in reality is quite different.

Contrary to what one might expect, UTM is not about "managing" traffic as the term is generally used and understood by the air traffic community, and it does not in any way interact with the existing air traffic control system. For that reason, it might be more accurate to think of UTM as a "traffic monitoring" capability specific to unmanned aviation.

The National Aeronautics and Space Administration (NASA) describes its vision of UTM as follows:

> Currently, there is no established infrastructure to enable and safely manage the widespread use of low-altitude airspace and UAS operations, regardless of the type of UAS. A UAS traffic management (UTM) system for low-altitude airspace may be needed, perhaps leveraging concepts from the system of roads, lanes, stop signs, rules and lights that govern vehicles on the ground today, whether the vehicles are driven by humans or are automated…
>
> While incorporating lessons learned from today's well-established air traffic management system, the UTM system would enable safe and efficient low-altitude airspace operations by providing services such as airspace design, corridors, dynamic geofencing, severe weather and wind avoidance, congestion management, terrain avoidance, route planning and re-routing, separation management, sequencing and spacing, and contingency management.
>
> One of the attributes of the UTM system is that it would not require human operators to monitor every vehicle continuously. The system could provide human managers with the data to make strategic decisions related to initiation, continuation, and termination of airspace operations. This approach would ensure that only authenticated UAS could operate in the airspace.
>
> NASA envisions concepts for two types of possible UTM systems. The first type would be a Portable UTM system, which would move between geographical areas and support operations such as precision agriculture and disaster relief. The second type of system would be a Persistent UTM system, which would support low-altitude operations and provide continuous coverage for a geographical area.[16]

In the conceptual diagram provided on their UTM home page (see below), it is clear that NASA's vision of the "need" to which they are responding essentially anticipates a band of airspace below 500 feet above ground level that would be the exclusive domain of unmanned aircraft. This would greatly reduce the consequences of failing to see and avoid in time to prevent a collision; the only other aircraft that should be in the vicinity would be similarly unoccupied.

[16] https://utm.arc.nasa.gov/.

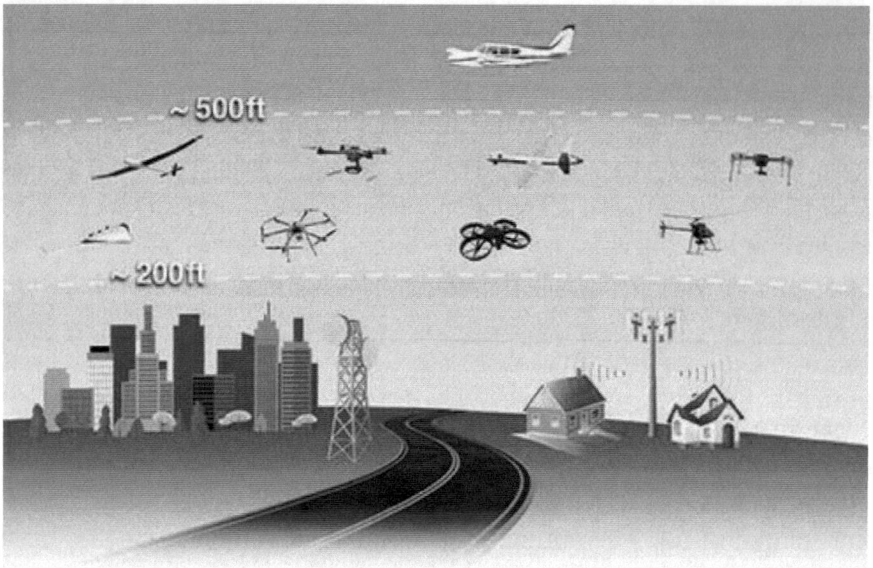

It is important to note that NASA's concept – while occasionally referring to "integration" elsewhere in program-related technical documents –is clearly directed toward airspace use that would be at its core *segregated* and totally devoted to supporting low-altitude UAS activity, in actual operation. While not congruent with the goals established by the FAA in its 2012 *Integration of Unmanned Aircraft Systems into the National Airspace System: Concept of Operations*,[17] UTM offers a realistic vision for traffic de-confliction that acknowledges the inherent risk in mixing manned and unmanned aircraft in the same volume of airspace, while simultaneously presenting a model for how to manage growth in the sector safely.

Sustainability of the Operational Approval Process

The sUAS regulatory construct described above, in combination with UAS limitations related to their see-and-avoid obligations, places a large administrative burden on both would-be commercial sUAS operators *and* the FAA, especially the Air Traffic Organization (ATO). While both the Unmanned Aircraft Systems Integration Office and ATO are required to review each waiver application, in practice, the burden is on the ATO and individual air traffic control facilities to approve each non-standard use of controlled airspace on a case-by-case basis.

Another equally important consequence of this new process is that it continues the pattern of requiring the FAA to make risk decisions regarding currently regulated activities involving non-conforming aircraft or operations. If observed in all particulars, Part 107 allows sUAS activity to take place in a manner and environment that minimizes risk to other users. As soon as any relief is granted from any part of that construct, risk in the NAS increases so as to require a separate, specific risk assessment of the non-conforming aspects of the proposed operation. COA applications are not assessed under the aegis of the FAA's "safety risk management" process, but for all practical purposes, every approval requires such an assessment.

[17] https://www.suasnews.com/wp-content/uploads/2012/10/FAA-UAS-Conops-Version-2-0-1.pdf.

If the unmanned aviation sector is to grow, the challenges their operations present from a purely regulatory perspective – especially in the context of their known limitations – must be approached differently. This may require flexibility in the application and enforcement of other long-standing regulatory components of the NAS.

However, it also requires the FAA to be permitted to let the pendulum swing back toward a neutral position between "promoting air commerce" and "promoting air safety" in future rulemaking. Lessons learned over time through accidents have shaped the current structure and operational framework for flying in the National Airspace System. These should not be erased simply because of the differences and limitations of unmanned aircraft. A middle ground needs to be sought.

Potential Registration and Equipage Rule Changes

As a first step (and a necessary prelude to rulemaking), FMRA will need to be revised to remove its prohibitions on rulemaking associated with model aircraft operations. The operation of small unmanned aircraft outside the existing regulatory structure is adding risk to the NAS itself. The model aviation community cannot police itself. So, hobbyists need to be required to accept greater responsibility for the safety of the finite resource that is the national airspace by becoming subject to more prescriptive, enforceable regulations.

Once legally permissible, the current requirement for universal *registration* of UAS[18] should go hand in hand with a requirement for universal *regulation* of UAS (exempting the proposed 'micro-UAS' class of extremely small unmanned aircraft). To some extent, this requirement is already in place. However, it needs to be modified and strengthened for two reasons:

1. Current provisions for identifying the registration number of an unmanned aircraft involved in a major accident (e.g., colliding with a manned aircraft) are unlikely to survive impact forces or post-crash fires. Solid, potentially survivable components, such as electric motors, battery packs or engine blocks, should be required to bear engraved, permanent registration information if there is to be any reasonable chance of tracing them back to their operator.
2. As with manned aircraft, registration information should be applied to the assignment of discrete Automatic Dependent Surveillance – Broadcast (ADS-B) identifiers to support universal equipage of all aircraft – manned and unmanned alike – with at least ADS-B Out. This would support surveillance in terminal airspace where low-flying small unmanned aircraft at low altitude may be in close proximity to arriving and departing aircraft at busy airports and bring the unmanned sector into closer compliance with "squawk and talk" requirements established in Part 91 for Class B and C airspace.

Despite protests to the contrary, additional safety-oriented regulation will in no way harm model aviation as a hobby, except to add some cost where model activity is to be conducted in close proximity to working airfields. Responsible modelers will be able to continue to operate much as they have over time, while *irresponsible* modelers could be subjected to civil or even criminal enforcement action as needed. The artificial distinction between commercial and "hobby" operations has proven meaningless in practice, and the steady growth of sUAS activity for all purposes mandates better regulation of it.

---

[18] https://www.faa.gov/licenses_certificates/aircraft_certification/aircraft_registry/UA/

Potential NAS Management Strategies

Once the FAA is free to act in ways that give equal consideration to all parties desiring or affected by the use of airspace by unmanned aircraft of all sizes and capabilities, it is likely that action in three separate domains will be required to safely accommodate (*not* "integrate") UAS in the NAS in the years ahead. In decreasing order of feasibility and increasing order of cost, they are:

1. Restructuring how airspace is used and by whom;
2. Re-thinking of the rules governing all aircraft operations in the NAS in consideration of UAS limitations; and
3. Investing in NAS infrastructure specific to the needs of UAS, but that results in a safer operational environment for all NAS stakeholders.

Re-imagining Airspace

The NAS has never been fixed in form and scope. It puts aircraft where they need to be to keep them safe and efficient, consistent with the needs of other NAS users. Now, unmanned aircraft challenge how the NAS itself currently is constituted in terms of airspace classification and apportionment.

Over time, countless changes to how the NAS is managed, controlled and protected have emerged based on the needs of both safety and commerce. For example, airspace needed for safe instrument approaches has come to be protected, charted and closely monitored for obstructions, and maximum airspeed and minimum vectoring altitude parameters have been established to support safe arrival, departure and close-in terminal operations.

The FAA cannot "create" airspace, despite frequent requests for them to do exactly that. "Corridors" and "tunnels" through controlled airspace have all been suggested to meet the needs of narrowly defined operating locations or to serve individual users. Beyond some of their underlying assumptions about airspace being flawed, such case-by-case proposals would be impractical to implement at a national level.

For that reason, some have suggested in effect "flipping the script": giving priority or even exclusivity to UAS in lower altitude bands and requiring other visual flight rules (VFR) operations to fly above them. Of course, there are several potential consequences of such a philosophical shift that could serve as lightning rods for resistance to such a notion, including:

- Adverse impact on manned aviation operations traditionally (or necessarily) conducted at low altitude, e.g., helicopter emergency medical service, public safety aircraft, aerial application operations, and electronic news-gathering.
- Reduced separation between VFR and instrument flight rules (IFR) operations where the former would wind up higher and closer to IFR Minimum Enroute Altitudes or Minimum Safe Altitudes.
- Increased likelihood that manned aircraft attempting to remain in visual meteorological conditions would wind up inadvertently entering instrument meteorological conditions.
- Reduced efficiency and increased cost for short-haul general aviation operations due to required climb above the altitudes typically used by such activities.

In effect, UAS are square pegs in round holes. They have operational needs and limitations worthy of consideration in the apportionment of space in which to operate, to the extent that such is possible without adversely affecting other stakeholders. Since they began flying in the NAS in significant numbers, managing UAS individually has consisted of rounding off the corners of the pegs. Perhaps it is time to begin making some of the "holes" permanently big enough to accommodate them, regardless of the UAS pegs' shape.

Reconsideration of Rules and Procedures

This second strategy relates to the first. Currently, the only process by which unmanned aircraft (other than sUAS) can gain access to the NAS is by exception. If a given operation requires relief from Part 107 provisions, or if Part 107 itself cannot be applied due to the maximum takeoff weight (MTOW) of the unmanned aircraft, all of the various non-conformities of a given UAS to existing Part 91 rules must be taken into account in granting a COA for specific operations.

There is growing acceptance of the proposition that some types of unmanned aircraft, or operations conducted by them, are poorly suited to full-up participation in the NAS. This is why sUAS operations are not considered either "VFR" or "IFR." Along these lines, existing requirements for transponders, two-way radio communications and the like may be impractical for a variety of reasons, or they may simply be unnecessary. But, they must be consciously addressed.

This suggests a need to rethink where UAS operations most commonly are conducted, as well as the Part 91 equipage requirements for access to certain classes of controlled airspace, from a regulatory perspective. Perhaps it would be appropriate and safe to excuse UAS from compliance with certain requirements in certain airspace; in those cases, they should be granted permanent regulatory relief from those requirements. On the other hand, it may be that, upon reflection, the growing amount of unmanned aircraft activity may not be tenable under some operational conditions unless they – as "aircraft" – are required to comply with existing rules.

Had Part 107 said, in effect, "all Part 91 rules apply to small unmanned aircraft, except as follows," the path to follow-on rulemaking for other classes of unmanned aircraft most likely would be much clearer. However, as is the case with other aspects of trying to make UAS operations in the NAS more routine, initial rulemaking and proposed follow-on rules are being accomplished on a patchwork basis in response to whichever part of the unmanned aviation sector currently is exerting the most political influence based on its real or potential contributions to the economy.

Whether addressing small UAS, large UAS or something in between, future rulemaking should keep four fundamental tenets of unmanned aviation as it relates to manned aviation firmly in mind:

1. **See and Avoid.** Any unmanned aircraft operation conducted beyond the visual line of sight of the pilot in command and/or visual or "electronic" observers (e.g., using UTM or some form of DAA) places all of the see-and-avoid burden on any manned pilots who might encounter it. Kinetic force and impact modeling, traffic density studies and the like are fine in the abstract, but they must lead to complete understanding and clear rules quantifying the risks that unmanned aircraft pose to manned aircraft, where one pilot can become a casualty as a result of another's actions. "See and avoid" is an established aspect of how

the NAS works, and unmanned aircraft systems must not be excused from conforming to it in some manner.

2. **Standardization.** There is no standardization among system types with respect to attributes of prime importance to those responsible for controlling or managing airspace (control link reliability, link failure response, on-board equipage for NAS operations, provisions for two-way radio communications, etc.). Without broadly applicable UAS certification and airworthiness standards aimed at reducing variable performance and unpredictable behavior, purely regulatory or procedure-based mitigations for their common limitations cannot succeed.

3. **Existing rules and procedures**. These should be applied to *all* aircraft uniformly unless the risk to human life can be minimized *and* the economic advantage is sufficient to justify their modification. However, by their very nature, unmanned aircraft are incapable of conforming to some rules. In such cases, requirements for alternate means of compliance must be explicitly and uniformly provided or, alternately, explicitly ruled out on a NAS-wide basis. Although virtually every Class B, C and D facility and airspace differs in some respects from all others of the same class, they are similar enough that general rules pertaining to all are possible. The same is true of unmanned aircraft systems; whether highly sophisticated or extremely simple, their commonalities are an adequate base upon which to design rules that can be broadly applied.

4. **Waivers.** Any time new rules or changes to existing rules are under discussion, the responsible parties should bear in mind the need to be as inclusive as possible for all modes of aviation, manned and unmanned alike. The ability to waive portions of a rule is important, but too much reliance upon that mechanism suggests a rule does not work as it should. In the unmanned aviation domain, some things shouldn't be waived under any circumstances and some should require uniform AMOCs if waived. Waivers that depart too far from Part 91 or 107 baselines should require approval at higher levels of the FAA than currently required due to the need for greater risk to be accepted on behalf of other NAS users.

NAS Infrastructure

There are various ways that infrastructure can be leveraged – or created outright – that enables UAS operations by greatly extending their range or by making the operations themselves safer for surrounding aircraft. These could include new or expanded means of supporting unmanned and manned pilots' situational awareness regarding their respective activities, regardless of whether they are in controlled or uncontrolled airspace. Four current or emerging technologies should be brought to bear on these needs.

1. **Air Traffic Control (ATC)-Pilot Communications:** The NAS Voice System (NVS) program could be leveraged to convey UAS pilots' ATC communications from certified, secured ground control stations directly into controllers' headsets via dedicated terrestrial use of the same Voice Over Internet Protocol (VOIP) approach already planned for NVS. This would eliminate the need for UAS operators to devote precious control link bandwidth

to support communications, and would preclude the loss of ATC communications in the event of control link failure by providing them a separate, dedicated path.

2. **UTM:** The UTM concept continues to evolve. It deserves funding with which to do so. However, the current concept is too narrowly drawn to benefit the NAS as a whole. UTM needs to build upon its current architecture by adding in interfaces with both NVS and the rest of the national command and control (C2) infrastructure so it truly supports *management* of unmanned aircraft as a component of overall NAS operations. UTM needs to provide unmanned pilots with situational awareness of surrounding manned operations, and vice versa. Ideally, UTM should become a full-up component of the Next Generation Air Transportation System (NextGen), and would expand on the current practice of using civil and military digital air surveillance radar systems to provide a standard system for ground-based detect and avoid capability for UAS too small to accommodate on-board DAA systems.

3. **ADS-B:** ATC and manned aircraft need to be aware of the presence of unmanned aircraft, especially in congested airspace. Although both sUAS manufacturers and operators resist the notion that the current Part 91 requirements for ADS-B Out by 2020 should apply to them, if small unmanned aircraft cannot be reliably contained to the "below 400 feet" stratum their presence must be known to all other aircraft in their vicinity. There would be a sizeable bill to pay for adding such a capability to unmanned aircraft, and a means for allocating aircraft identifiers to the requisite number of ADS-B sets may not yet exist, but the emergence of ADS-B as the NextGen surveillance system of choice means UAS *must* participate, or they will be effectively invisible to other users and controlling agencies alike.

4. **Command and Control (C2) Network**: Despite the ever-growing collection of business cases for unmanned aviation, it is possible to categorize them as working within one of four types of operational areas, defined in terms of where their ground control station or stations are located and the "reach" of their supporting C2 architecture:
    - Close-in/within visual line of sight (VLOS) only.
    - Defined area beyond visual line of sight (BVLOS).
    - Regional (a larger, defined BVLOS area requiring satellite or terrestrial support to maintain control link continuity).
    - Point-to-point/long-distance (same as above, but requiring sequential changes in the source of C2 connectivity as the flight progresses).

    None of the above are explicitly bound to a given altitude domain, and none can be restricted to a single class of airspace. As such, a class-based "airspace" solution supporting any of them is not feasible. At the same time, the bulk of unmanned aircraft operations (except those that are purely VLOS) require both VLOS and BVLOS C2 capabilities with which to operate.

    RTCA is already suggesting that the FAA needs to get into the "frequency allocation" business to allow UAS to operate in the various regimes listed above; the next logical step would be for a conscious decision to be made for the FAA to build a national backbone for

UAS C2 management itself. This would be analogous to the system of ground stations required for national ADS-B deployment, or to the legacy network of surface-based navigational aids.

In short, unmanned aviation entails new risks to manned aviation that the current air traffic control system cannot mitigate or manage without enlisting new technologies to the task. As a permanent part of the operational landscape, it will require permanent investment and permanent solutions.

Conclusion

UAS operations are unlike manned operations in a number of respects. Most notably, they offer a far less expensive means of obtaining a desirable aerial perspective. Manufacturers and operators have little interest in seeing the cost of UAS operations driven up by other users, but they must bear some responsibility for keeping the existing system safe.

Given the challenge that UAS presents to the current regulatory environment, it may be wise to consider the words of futurist Alvin Toffler from almost a half-century ago, in his seminal work *Future Shock*:

> With such large sums at stake, one would think that governments would plan their technological development carefully, relating it to broad social goals, and insisting on strict accountability. Nothing could be more mistaken.
>
> "No one – not even the most brilliant scientist alive today – really knows where science is taking us," says Ralph Lapp, himself a scientist-turned-writer. "We are aboard a train which is gathering speed, racing down a track on which there are an unknown number of switches leading to unknown destinations... Most of society is in the caboose looking backward."[19]

"Future shock is a time phenomenon, a product of the greatly accelerated rate of change in society. It arises from the superimposition of a new culture on an old one."[20]

The disruptive power of unmanned aviation has yet to be fully realized or managed. However, it is becoming evident that the unmanned aviation sector must receive a reasonable amount of priority in future airspace and air traffic management planning and programming, in the name of both air commerce *and* air safety. The challenge now is to acknowledge the trends and take concrete action in response to them, sooner rather than later.

In addition to being an active individual member of ATCA, Tom Farrier is the Chair of the International Society of Air Safety Investigators (ISASI) Unmanned Aircraft Systems Working Group, and is a long-standing member of the Flight Safety Foundation and the American Society of Safety Engineers. He is a former National Safety Coordinator for the National Air Traffic Controllers Association and a former Director of Safety for the Air Transport Association (now "Airlines for America"). He has provided contract support to both the Department of Defense and the FAA on UAS safety matters since 2006. The views expressed in this paper are the author's and do not reflect official positions of ISASI, or of JMA Solutions, LLC or its clients.

[19] Alvin Toffler, *Future Shock* (New York: Bantam Books, 1970), p. 431.

[20] *Ibid*, p. 11.

Artificial Intelligence in ATM

Marco Rueckert
Searidge Technologies
19 Camelot Drive, Ottawa, Ontario, Canada, K2G 5W6
marco@searidgetech.com

Abstract

Due to the recent advancements in technology, Artificial Intelligence (AI) is able to solve real world problems today. Many industries are already benefiting from AI. This paper aims to present how the Air Traffic Management (ATM) industry can benefit from AI and related technologies. The paper describes how AI has the potential to shorten the development time of ATM software systems for which traditionally human experts design a system and create a sophisticated model of the real world problem/challenge. The potential to shorten the development time of such systems is illustrated using an early go-around detection system as an example. Finally, the paper presents how additional areas of the ATM industry can benefit from AI.

Introduction

Artificial Intelligence (AI) has received a lot of attention in the media in the recent past and has been used in many industries. The applications for which AI has been utilized are diverse, ranging from self-driving cars to stock trading. Artificial Neural Networks (ANNs) are one technology which has lately been used in many AI applications and projects. ANNs are networks of artificial neurons, which were designed based on a model of the biological neuron [1, p. 6]. These artificial neurons are grouped together in layers to form ANNs. Artificial Neural Networks are ideal for pattern recognition tasks such as image classification and object detection in images. ANNs learn to recognize specific patterns in data through a training process. One common training method is called *supervised learning* during which the ANN processes annotated training data. Annotated training data consists of samples of input data which have been associated with the expected network output. In other words, the desired output values are fed into the system along with the sample data. The weights of the connections between the neurons in the ANN are adjusted during the training process in order to minimize the error between the actual and desired network output. The most commonly used algorithm to adjust the weights between the neurons is called *backpropagation* [1, p. 10]. The ANN learns a function approximation of the real world problem it was designed to model. This function approximation enables the ANN to not only achieve the desired output to a certain degree of accuracy for the training examples, but also for never before processed input data [2].

Although the invention of the ANN technology and the modern backpropagation learning algorithm date back to 1943 [1, p. 6] and 1986 [3] respectively, only the recent advancements in processing hardware have made it feasible to use ANNs to solve complex real world problems such as self-driving cars. General-Purpose Computing on Graphics Processing Units (GPGPU) refers to the utilization of Graphics Processing Units (GPUs) to solve computational intensive problems. GPUs, commonly referred to as graphics cards, were traditionally only used for computer graphics. GPUs accelerate ANNs through massive parallelization. This means that graphics cards are able to perform the computations required for ANNs in parallel, resulting in a significant execution speed improvement compared to the processing hardware used before GPGPU. This improvement has enabled many research projects and technology solutions which were previously infeasible. For the purposes of this paper, all references to AI refer to the ANN technology. ANNs can be used for computations not related to AI, however this paper only discusses the application of ANNs to AI. Other technologies for AI exist, but these are not explored in this paper.

Artificial Intelligence and ATM

ANNs have proven to be a very powerful technology which has been applied in many industries to solve challenges such as Natural Language Processing (NLP) and language translation. However Air Traffic Management (ATM) is not just a single problem or challenge. It is a complex structure of integrated technologies, systems, regulations, and stakeholders. ATM as an industry is also safety-critical and risk-averse because any mistake could have severe consequences for aircraft passengers, crew members, and also people on the ground. The impact of an AI program mistranslating part of a sentence in social media can not really be compared to clearing an aircraft to land in unsafe conditions. Therefore, the question arises: How much can AI really improve the ATM industry? Due to the complexity and inter-connectivity of ATM systems, it would be naive to think that AI could replace or automate entire ATM systems or perform ATC tasks at a level superior to human experts in the immediate future. However, AI does not have to take over entire ATM operations to yield some benefits for the industry. Applying AI and technologies such as ANNs could improve the operation of single, smaller parts of larger ATM systems.

A paper by Julian et al. demonstrates how a neural network can be used for the next generation of collision avoidance systems called ACAS X. The decision logic for ACAS X systems is commonly stored as a large numeric table, but the ANN was able to approximate the numeric table as a complex non-linear function. This means that the ANN can approximate the decision logic represented in the entire numeric table with just the parameters of the network. This resulted in a compression of the information required to make decisions about the optimal aircraft maneuvers by a factor of 1000. The authors of the paper found that not only did the ANN compress the decision logic data significantly, but the network's performance actually surpassed the performance of the numeric table [4].

There are many more areas of the ATM industry which could potentially benefit from the use of AI. Nevertheless, it is also important to note that AI technology is not always the best choice to solve a problem. ANNs are particularly good at detecting patterns in data. A large amount of data is required to train the ANN in order for it to reach a good function representation of the real world problem it is modelling. Therefore, if training data is either difficult to collect or non-existent, then an ANN is most likely not well suited to solve the problem at hand. Traditional[1] software programs are also better suited to challenges for which a good model of the real world problem already exists and can be easily implemented in software, such as aircraft separation and velocity calculations. AI technology should not be used in favour of traditional software programs for these kinds of problems.

Go-Around Detection

The advantages of using an AI system instead of a traditional software system become apparent when

1 Traditional software system/program in the context of this paper refers to a non-AI program based on, for example, decision trees. This non-AI program is written with programming paradigms such as procedural and object-oriented programming.

the development of a go-around detection software system is investigated. A go-around occurs when the pilot of an aircraft decides to abort the final approach or landing of the aircraft. To decide whether or not to initiate a go-around, the pilot assesses the risk associated with the continuation of the final approach and landing attempt and the risk associated with a go-around. The pilot then chooses the safest option and follows Standard Operating Procedures. Should the pilot decide to conduct a go-around, he/she then decides whether to conduct another final approach and landing attempt at the same airport or to attempt to land at another airport. There are a number of reasons why the pilot of an aircraft may decide to initiate a go-around. These reasons include deviation from the nominal flight path and challenging weather conditions. An Air Traffic Control Officer (ATCO) may also instruct the pilot to conduct a go-around for a number of operational reasons.

Go-arounds are an unwelcome surprise for aircraft passengers and an inconvenience for ATM personnel, but a go-around by itself is a safe aircraft maneuver and is well defined in procedural guidelines for each airport. Even so, if the ATCO controlling the aircraft which is conducting a go-around is not aware that the aircraft is conducting a go-around, a safety risk can arise. This is especially true for large and busy airports with multiple dependent runways. The workload for pilots during the final approach and landing of an aircraft is high and therefore occasionally pilots inform their ATCO of their decision to conduct a go-around only after they have initiated the go-around. The primary concern of the pilot is the safety of the aircraft and the on-board passengers and not the communication with the ATCO. If the ATCOs are not aware of the go-around of a specific aircraft, they may make decisions based on the assumption that the aircraft actually landed. For example, an ATCO may clear another aircraft on an adjacent and dependent runway for take-off during the landing of the aircraft which is conducting a go-around. This scenario can result in dangerous situations in which the horizontal and vertical separation requirements between the aircraft are not satisfied.

To help avoid these dangerous situations LVNL (Luchtverkeersleiding Nederland), the Dutch Air Navigation Service Provider (ANSP), has developed their Go-Around Detection System (GARDS). The system developed by LVNL uses an algorithm to detect go-arounds based on surveillance data. The system warns the ATCOs with visual and audio warning reports if the system has detected a go-around. LVNL has optimized the system to reduce the number of false alarms to a minimum [5].

Go-Around Detection System Development

Traditionally, when a new software system such as a go-around detection system is to be developed, human experts including ATC controllers, ATM consultants, and software engineers have to work together collaboratively in order to understand the problem and come up with a good model of the real world problem. For this, the human experts will investigate which data sources are available and applicable for the system, what the expected output of the system is, and how the real world problem can be modelled in such a way as to produce the correct output for each input or combination of inputs. In the case of a go-around detection system, the human experts would have to first determine at which point during the final approach of an aircraft the go-around needs to be detected. If the go-around should only be detected once the aircraft has either landed or has gone around, it would probably be enough to only build a model based on surveillance data of the aircraft (latitude and longitude, altitude, velocity, and heading of the aircraft). However if the go-around should be detected/predicted as early as possible during the final approach of the aircraft, it would probably be not enough to build a model only based on surveillance data. Additional factors such as weather data (wind speed, wind direction, and wind gust information), time of day, time of year, and aircraft type[2], which also influence the probability of an aircraft having to initiate a go-around, also have to be considered. The output of the go-around detection system is rather simple[3]. The system should just output one value: the probability of the analyzed aircraft having to initiate a go-around. From another perspective, this value represents the confidence of the system that the specific aircraft will conduct a go-around. This decimal value would range from 0 to 1, where 0 would represent total confidence that the aircraft will not conduct a go-around and 1 would represent total confidence that the aircraft will conduct a go-around. The described high-level design of the go-around detection system is shown in figure 1.

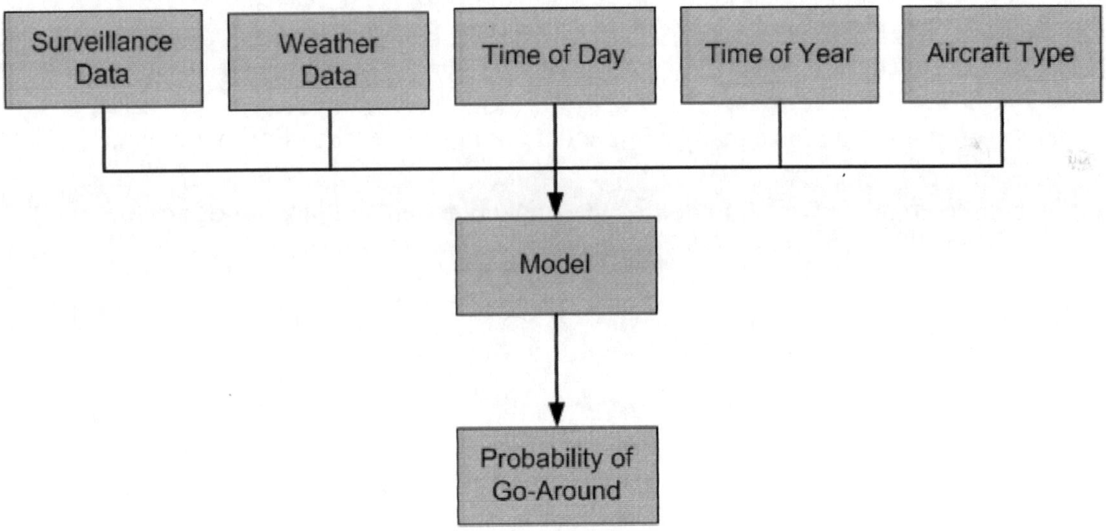

Figure 1: High-Level Design of the Go-Around Detection System

2 Time of day, time of year, and aircraft type are used just as examples for other factors influencing the probability of an aircraft having to conduct a go-around. Testing would have to be performed to select these factors.
3 The described go-around detection system output value is not the final output presented to the ATCOs. The system output value would be used to format the final output presented to the ATCOs.

In the case of early go-around detection[4], the system would have to compare the flight path of the aircraft to the nominal flight path and then could predict a go-around with a certain degree of confidence at various stages during the final approach of the aircraft based on the surveillance data and other factors previously mentioned. To build this system, the human experts would examine sample data for a number of final approaches and landings of aircraft. This sample data should be spread out over different weather and traffic scenarios and should involve many different types of aircraft. One possibility to build a model for the early go-around detection system is to analyze the data from the various input sources at predefined positions, i.e. threshold distances from the runway at the airport. This would allow the system to take measurements from all input sources at these threshold distances from the runway as shown in figure 2 below. These measurements could then be analyzed over the entire sample input data space in order to understand how certain measurements at certain distances from the runway affect the probability of an aircraft having to initiate a go-around. Essentially, the human experts would try to analyze many sample data points in order to find patterns in the data which would allow them to come up with a model for the final approach and landing of aircraft.

4 In the following discussion, this paper investigates the early go-around detection system only.

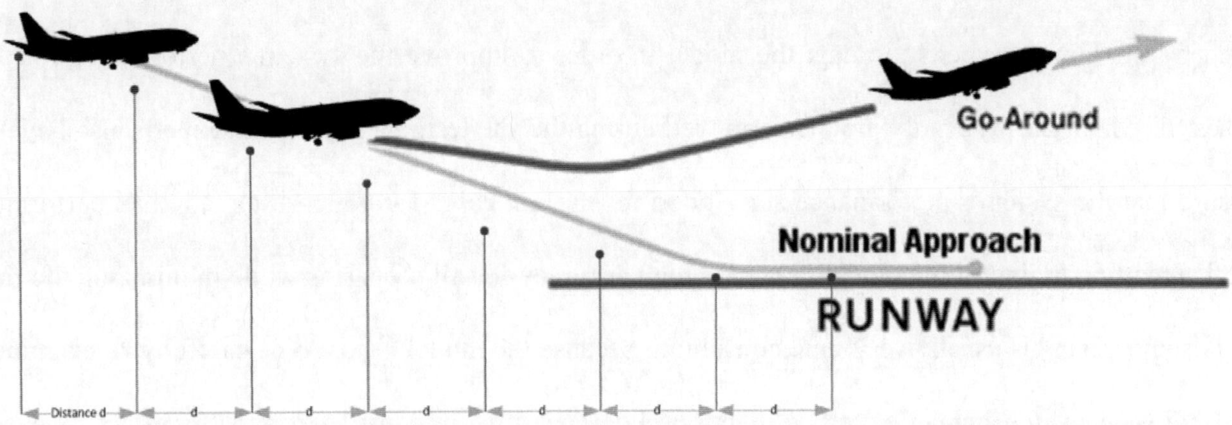

Figure 2: Measurement Distances of the Go-Around Detection System

Once the experts have come up with a model which accurately classifies the sample input data as either go-arounds or landings, this model would be implemented and then tested on live data at the airport for which the system was designed. The test cases would, for example, test that the go-around system correctly identifies 95 out of 100 go-around maneuvers as go-arounds before the aircraft has reached a specific threshold distance towards the runway. Another test would measure the false alarm rate of the system and ensure it was below a predetermined threshold rate in order to ensure that ATCOs could confidently rely on the system's output. It is expected that the first implementation of the software would not correctly classify all of the aircraft maneuvers as go-arounds or normal landings in the real world because the model was based on sample data which most likely did not include all possibilities. Human decision making also affects whether the pilot of an aircraft conducts a normal landing or a go-around. Where all the conditions indicate that a go-around is the appropriate course of action, most pilots would be expected to choose to conduct a go-around, while human factors may lead another pilot to continue with the landing of the aircraft. In order to improve the system, feedback for the human system development experts would be collected during the testing on site. This feedback would then be used by the human experts to adapt the model in order to improve the system's performance for the cases for which the system did not perform well during the site testing. The human experts also have to ensure that the system's performance stays good for the test cases for which it already had performed well. This fine tuning of the model to ensure high accuracy for all scenarios while minimizing the rate of false positives[5] is usually very time consuming because the model needs to be carefully re-examined and adjusted by the human experts. All possibilities need to be considered and the effect of every change in the model on the overall performance of the system needs to be evaluated and validated.

In the case of an AI-based early go-around detection system, the development process of the model of the real world problem is very different from the process used for traditional software systems outlined above. As described in the introduction of this paper, we will consider only the ANN technology for AI systems in the paper. ANNs learn the model of the real world problem through a training process. There is no need for human experts to understand all of the details of the problem and come up with a model. For the AI system, the input data sources would be the same as for the traditional software system described above. Namely, the inputs for the ANN would be surveillance data, weather data, and any other relevant and available data such as the time of day, time of year, and aircraft type. Similarly to the model designed by the human experts discussed above, one possibility to design the ANN for the early go-around detection system is for it to take measurements from all input data sources at predefined threshold distances from the runway. The measurements would then translate to clusters of input neurons in the ANN. There would be one cluster for each measurement distance from the runway and

5 In the case of go-around detection, a false positive means the system alerting the ATCO of a go-around although the airplane has landed normally or will land normally.

there would be one input neuron for each input data value such as aircraft altitude, heading, wind speed, and wind direction. These input neuron clusters are shown in figure 3 below, but for illustration purposes each input neuron cluster is just shown as one input neuron. Figure 3 also shows a simplified[6] ANN network architecture.

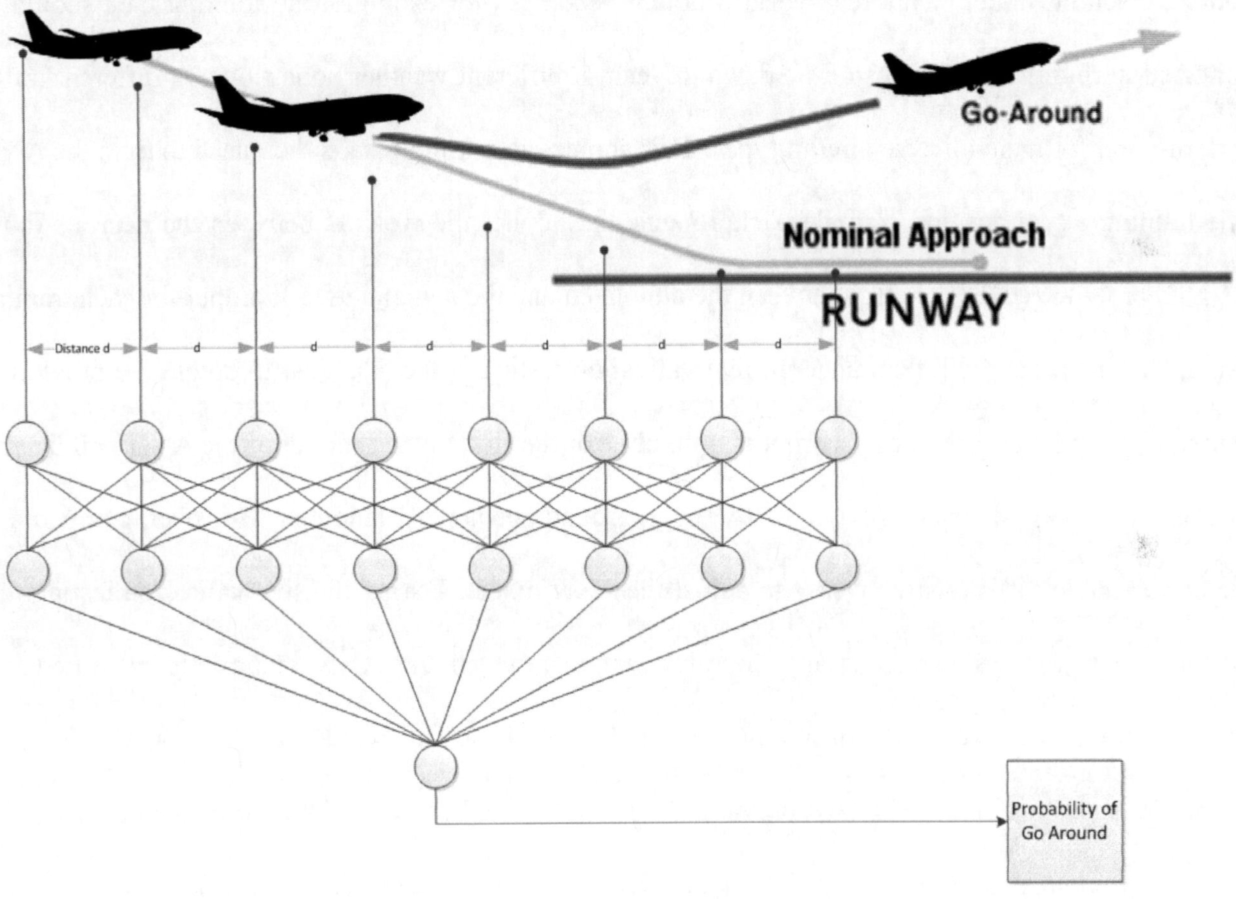

Figure 3: Sample ANN Architecture of the AI Go-Around Detection System

To create annotated training data for the training (supervised learning) of the ANN, a human only needs

6 The ANN architecture was simplified for illustration purposes. In reality the architecture would be more complex.

to annotate the live data from all input sources. The human annotates each training example (aircraft maneuver) as either a normal landing or go-around and records the data from all input sources. The input data recording along with the annotated classification (go-around or landing) forms one training data sample for the ANN. The annotations should include many different aircraft maneuvers in many different scenarios. The greater the number of different cases of input data provided to the ANN is, the better the learnt model of the real world problem becomes. For example, the training data should be collected during different times of the day and year, in different weather conditions, in different traffic patterns, and for many diverse aircraft types. This annotated training data is then used to train the ANN. The training algorithm, for example backpropagation, adjusts the weights between the neurons in the ANN in order to reduce the error between the annotated and the actual network output for each training example. The system will then undergo live validation testing at the site. Just like with the previously described traditional software system, it is unlikely that the first trained model of the ANN will be able to classify all aircraft maneuvers correctly as go-arounds or normal landings. However, compared to the expert model, it is relatively easy to adjust the ANN model. During the live validation testing, the human tester just has to record any aircraft maneuver which the ANN incorrectly classified and annotate it with the correct classification (go-around or landing). This would be as easy as watching the air traffic from the airport tower and pressing either an "accept" or "reject" button on a software which records the input data. In this way, the human tester can easily record feedback for the next training iteration of the ANN. The ANN training algorithm then uses this new training data to fine tune the model to correctly classify the maneuvers recorded during the testing. By not only using the new training data, but also the previously learnt data, the ANN is fine tuned in a way to preserve the previously learnt information. This way, the ANN can learn to correctly process the new training data, while still correctly processing the previously learnt training data. The development process of the model for the AI-based early go-around detection system is illustrated in figure 4 below[7].

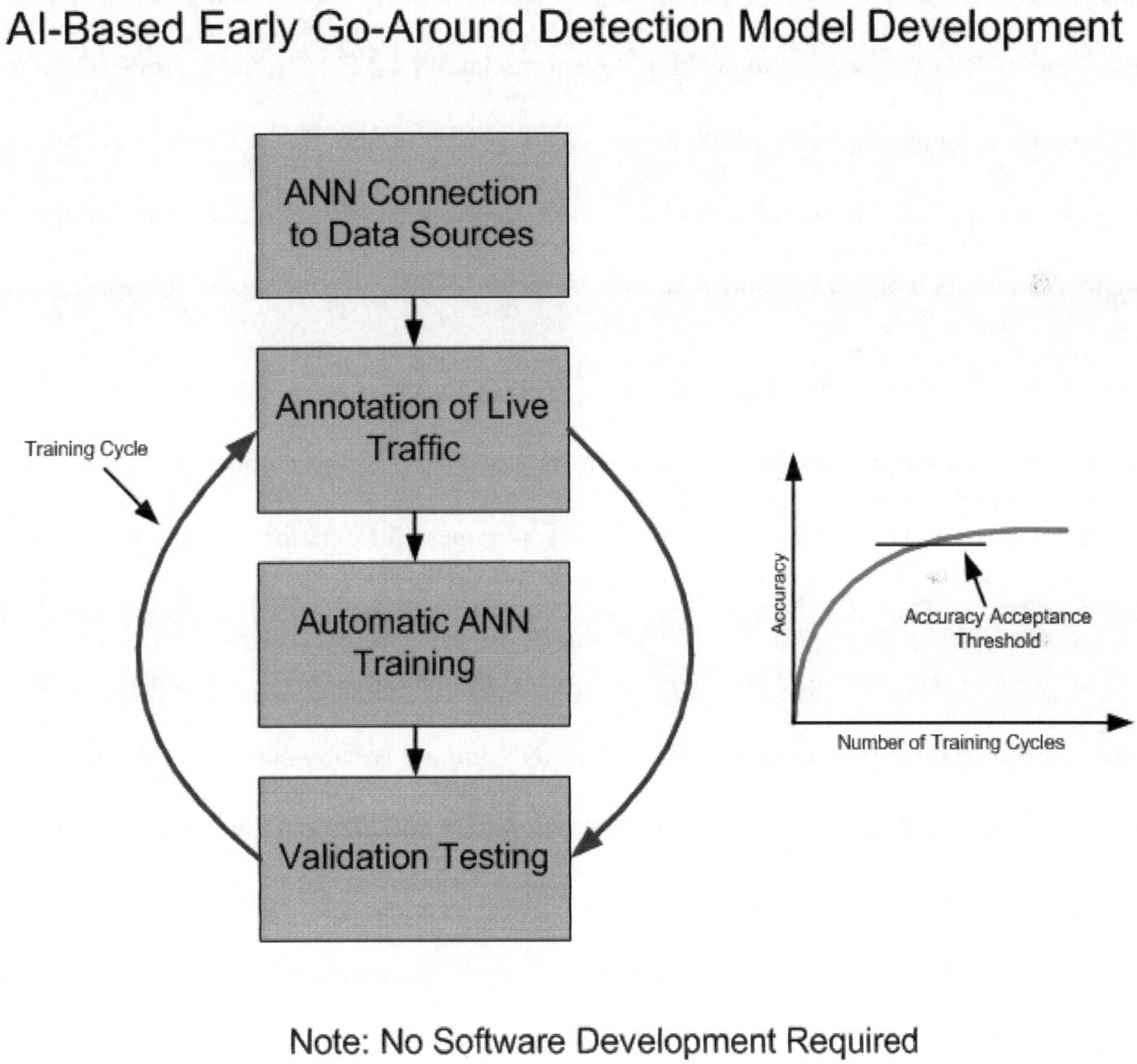

Figure 4: AI-Based Early Go-Around Detection Model Development

7 The accuracy versus number of training cycles graph shown in figure 4 is meant just for illustration purposes.

Comparing the work which is required to create a sophisticated model of the real world problem for a traditional software systems to the work which is required to train an ANN, it becomes clear that the development time for a system such as the early go-around detection system can be shortened significantly by employing an ANN. Additionally, the level of education and knowledge required to annotate training samples for an ANN, and then train it, is much lower compared to the level of education and knowledge required to develop a good model of an ATM problem. Furthermore, there is usually a team of human experts needed to model an ATM problem accurately, while it would only require a trained specialist to train the ANN. In short, using the ANN technology has the potential to save many man-hours in the development process of ATM systems, depending on the problem.

Since the ANN learns the model of the real world problem using an algorithm, the technology is also able to find patterns which humans would have never considered. If the number of input sources is as large as for the go-around detection system, it is sometimes difficult for humans to find the most relevant patterns in the input data space. In comparison, the ANN will just systematically iterate through all the training data over and over again until the model error is reduced to an acceptable minimum value. Through this training process, the ANN implicitly finds the most relevant patterns in the input data space. This advantage is described in an article in Scientific American outlining how AI is used for research into earthquake warning systems [6]. The article states that the AI was able to find patterns in the raw data collected for earthquakes which the researches had never found because they were not investigating the raw data, but rather processed seismic data [6].

AI Potential

AI does not only have the potential to drastically reduce the development time of ATM software systems, but it also provides the opportunity to solve problems which were previously believed to be unsolvable. AI gives ATM system developers a whole new perspective and a wide range of new technologies to develop ATM systems which have the potential capability to tackle some of the most challenging problems in the industry. Any system that analyzes large amounts of data to find certain patterns in the data can benefit from AI. It is rather unnatural to think about ATM problems in terms of data patterns. However, the amount of data in the ATM industry is vast and very diverse. This data diversity is also an advantage for ANNs. Training ANNs with many different scenarios, e.g. weather patterns, aircraft types, and traffic patterns, inherently makes the ANNs resistant to noisy data [1, p. 35]. Diverse training data also helps the ANN learn an abstract model of the real world problem. This means that the ANN cannot only correctly process the training input data samples, but also learns how to process never before processed input data [2]. This gives ANN models another advantage over models created by human experts, namely to correctly handle outlier data. In the case of the early go-around detection system, this would be, for example, a rare wind pattern which drastically increases the probability of a certain type of aircraft having to conduct a go-around. A properly trained ANN would then correctly classify the affected aircraft maneuvers as either a go-around or a normal landing based on the abstract model the ANN has learnt.

As previously discussed, AI has the potential to find patterns in data which a human would have never thought to investigate. Therefore, AI has the potential to improve ATM operations in ways humans have not yet thought of. For example, AI could enhance Airport Collaborative Decision Making (A-CDM) by processing data from multiple sources such as ADS-B, METAR and airline scheduling tools. The AI A-CDM system could assign gates to planes automatically in such a way as to reduce the taxi distance and time for aircraft between arrival and departure based on for example the current and learnt traffic and weather patterns. The optimal gate assignments would result in lower fuel consumption and emissions for taxiing aircraft. Another example would be an AI system that provides suggestions to ATCOs about the optimal flight level for each flight within the US National Airspace System (NAS) in order to minimize en-route fuel consumption. The same system could be used to avoid congestion and delays at airports by consistently providing optimal suggestions to the ATCOs and thereby reducing service level variations.

Conclusion

Due to the recent advancements in technology, AI is able to solve complex real world problems today. Many industries are already being re-shaped by AI, and the ATM industry has a large potential to benefit from AI. However, it has to be made clear that AI technology is currently still far away from the AI capabilities shown in many science-fiction movies. Due to the complexity, risk-averse nature, and inter-connectivity of the ATM industry, AI will not be able to replace entire ATM systems or even human experts/operators any time soon. This paper has shown that the immediate potential of AI in the ATM industry, is in providing ATM system developers with a new set of technologies to build assistance systems for human operators and provide solutions to ATM problems that were previously thought to be unsolvable. The case study of the early go-around detection system showed that AI technology can shorten the development time of ATM software systems and increase the system's resilience to noisy and outlier data. This will result in more accurate systems with lower error rates that can get deployed faster than ever before. Overall, AI has the potential to improve the quality of the services provided by the ATM industry.

References

[1] Christopher MacLeod, "An Introduction to Practical Neural Networks and Genetic Algorithms For Engineers and Scientists"

[2] Andrew J.R. Simpson, "Abstract Learning via Demodulation in a Deep Neural Network", ArXiv e-prints, eprint arXiv:1502.04042, Feb. 2015

[3] David E. Rumelhart et al., "Learning representations by back-propagating errors", Nature Publishing Group, 1986

[4] Kyle D. Julian et al., "Policy Compression for Aircraft Collision Avoidance Systems", IEEE/AIAA 35th Digital Avionics Systems Conference (DASC), Sacramento, CA, 2016

[5] CANSO, the Civil Air Navigation Services Organisation, "Go-Around Detection System - Working with a safety net" in *Airspace - Journal of the Civil Air Navigation Services Organisation Issue 34 Quarter 3 2016*, Hoofddorp, The Netherlands: CANSO, the Civil Air Navigation Services Organisation, 2016, pp. 30-31

[6] Annie Sneed, "Can Artificial Intelligence Predict Earthquakes?", Scientific American e-print, Feb. 2017

An Analysis of En Route Wake Turbulence Behavior based on In-Flight Measurements

Rocio Frej Vitalle, CSSI Inc.

Abstract

The recent incident involving the encounter of an Airbus A380 wake vortex by a Challenger 604 has drawn attention to the potential hazard of wake vortices en route. To broaden the understanding of the risks associated with en route vortex systems, the FAA in collaboration with the National Research Council of Canada (NRC) has collected high altitude wake turbulence data. An analysis of wake vortex fields based on such in-flight measurements has shown that wake turbulence may exist at distances well beyond minimum separation. The strength of these vortices, as well as transport and decay rates, was found to be highly dependent on atmospheric conditions. A study of historical weather data provided herein suggests that the likelihood for atmospheric conditions that trigger a wake encounter increases as longitudinal separations are reduced. Implementing wake mitigation strategies may allow increasing NAS capacity without compromising safety. Along track separation increases, altitude changes or lateral offsets are discussed as potential mitigations that may be useful in avoiding wake turbulence.

Introduction

Reducing en route longitudinal separation is a natural step in increasing the capacity of the National Airspace System (NAS). However, wake turbulence upsets are a limiting factor when considering the implications of new separation standards. Wake upsets continue to be reported at in-trail distances well beyond the minimum separation. Maintaining or improving current levels of safety while increasing capacity requires an assessment of the behavior and transport of wake vortex fields. Conversely, a lack of measured data at high altitudes has limited the understanding of aircraft vortex behavior during the en-route phase of flight. To address this data gap, the Federal Aviation Administration (FAA), in collaboration with the National Research Council of Canada (NRC) has collected high altitude aircraft vortex data. The data was collected by directly measuring wake vortex wind fields when in-trail of commercial transport category aircraft. This research presents a study of the atmospheric conditions that increase the potential for an encounter based on transport and behavior of vortex fields. In scenarios where wake turbulence avoidance is necessary, mitigation options are discussed such as altitude changes, vectors off route or procedural maneuvers in oceanic airspace. A need exists for a more complete understanding of wake turbulence behavior in the upper atmosphere and also for mitigation procedures that offer the optimum efficiency and enable wake safe head-on passing and in-trail altitude changes.

In-Flight Vortex Wind Field Measurements

The high altitude wake measurements were conducted by the NRC using a retrofitted CT-133 aircraft equipped with research grade sensors. Each of the data collection flights were conducted behind large, heavy or super heavy weight class transport aircraft in revenue service. The flights were coordinated with ATC to ensure minimum separation at the start of the data collection. As the run progressed, the separation between the CT-133 and the generating aircraft increased as an artifact of the speed differential between the two aircraft. To visually identify the location of the vortex pairs and ensure that measurements were obtained by intercepting the vortices, the condensate trail was used as a marker. Analysis of these vortex systems has enabled an estimation of their overall circulation strength, decay rate and descent.

A sample set of vortex intercepts is presented in **Figure 1** with corresponding in-flight cockpit imagery from the measuring aircraft. A Google Earth depiction of the generator and measuring aircraft paths along with the vortex system trajectory was produced using WakeWISE, a 4-D probabilistic modeling tool developed by the National Center of Excellence for Aviation Operations Research (NEXTOR). WakeWISE utilizes the position data for both aircraft along with the measured atmospheric parameters and generates a probabilistic estimate of the wake vortex transport. Because the model also accounts aircraft characteristics, it is able to estimate the strength of the vortex pair. In this example, vortex wind fields were measured nearly 20 nautical miles in-trail of the generating aircraft.

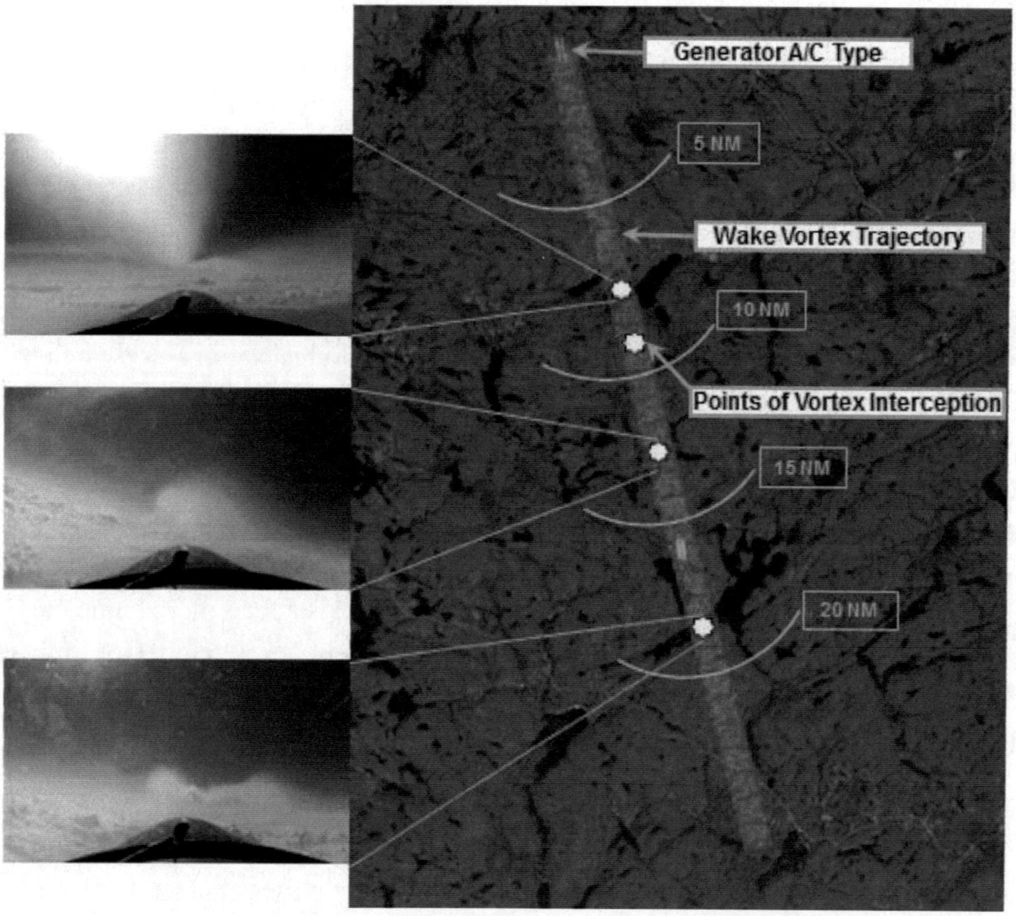

Figure 1: In-Trail Wake Vortex Trajectory

Wake Vortex Characteristics and Behavior

Wake vortices are a result of the pressure difference between the top and bottom surfaces of the aircraft wing. The higher pressure air from beneath the wing flows around the wingtips, and due to the forward movement of the aircraft, forms two counter-rotating vortices that are transported with the air mass. By decomposing the measured winds on the cross-sectional plane of the vortex system, it is possible to define the vortex and analyze its behavior and strength. Using the in-flight data collected by the NRC, vortex crossings were defined by rapid changes to winds in both the vertical Z-axis and the lateral Y-axis, as diagrammed in **Figure 2**.

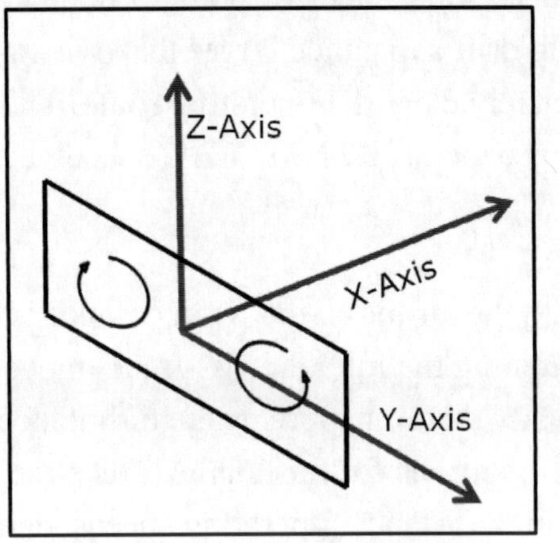

Figure 2: Axis Orientation

In **Figure 3**, the distinct shapes of vertical winds that define a vortex system are illustrated. The left vertical wind plot is an example of winds measured during a horizontal traverse of the left and right wing tip vortices, while the right vertical wind plot is typical of a single vortex core.

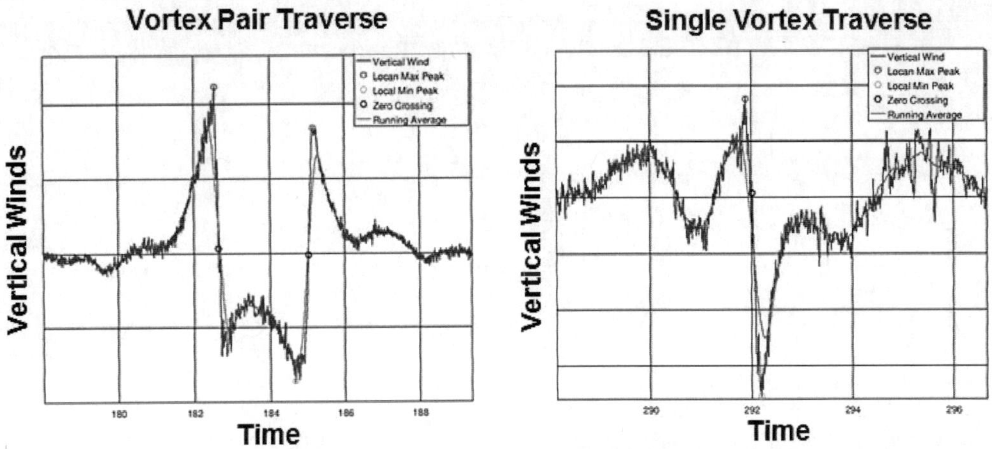

Figure 3: Vertical Wind Component Plots

The vortex pair is transported with the surrounding air mass and simultaneously begins to descend below the altitude of the generating aircraft's trajectory. Typically, wake vortices decay over a period of 2-3 minutes to near "background" turbulence levels. However, the analysis of high altitude wake vortex data provided by NRC has demonstrated that, in some instances, significantly strong vortex cores exist at distances in trail well beyond current separation minimums. Descent and decay rates can also vary, depending on atmospheric conditions, making precise predictability difficult.

Wake Vortex Transport

Under most wind conditions the vortex windfield is drifted with the air mass away from the operating route before a trailing aircraft reaches a point of interception. For the basis of this analysis, the wake is required to drift a distance larger than the pair-wise cross-track navigational positional uncertainty to completely clear the operating route. This distance is notionally assigned for the purposes of this paper as 0.3NM. Wind speed variability, which is also a factor, is not included in this estimate.

Because the wake is drifted with the air mass, the wind component perpendicular to the operating route is the key factor in determining the rate of cross-track transport. An analysis was conducted to calculate the crosswind velocity that transports wake turbulence away from a primary navigation route. The specific conditions for this analysis take the following variables into consideration: pair-wise separation distance, operating speeds, and a notional 0.3NM cross-track drift distance.

In **Figure 4**, the minimum threshold crosswind velocity to transport the wake turbulence away from the operating route was calculated for a set of ground speeds, varying from 300 to 450 KTS, and distances in trail between 3 and 20 NM.

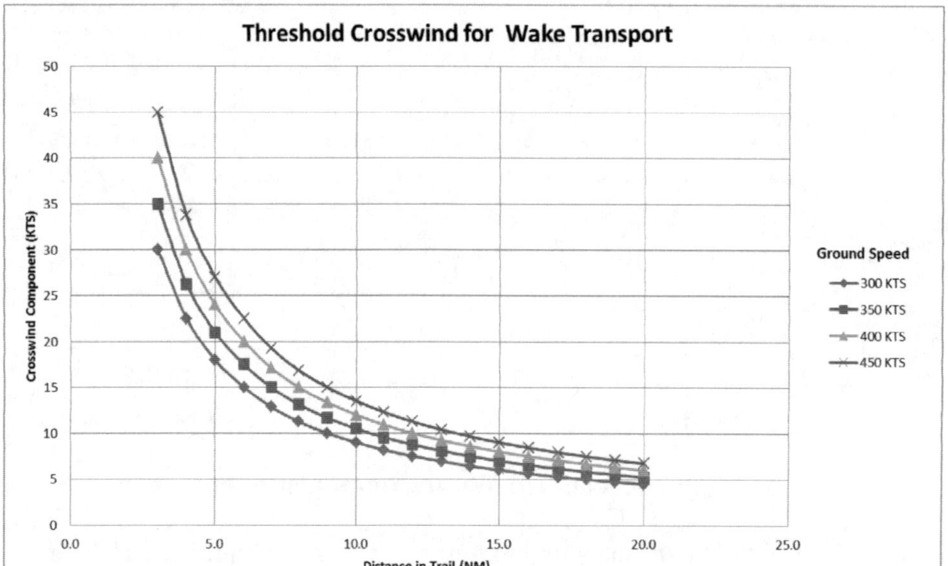

Figure 4: Threshold Crosswind Analysis

While the likelihood of these wind conditions to exist in combination with decay and descend rates that result in an encounter is minimal, it can be observed that in general, as separation distances are reduced, a higher crosswind is required to transport the wake away from the operating route.

Historical Wind Frequency Analysis

An analysis was performed to assess cross-track winds on select high altitude jet routes. Historical wind reports were analyzed to assess the frequency of cross-wind velocities that would not transport the wake away from the operating route on the previously presented scenarios. The route segments used for analysis were selected between four city pairs that might be expected to absorb high volumes of air traffic and that were also located in geographically diverse areas of the Continental United States (CONUS). The selected routes were situated in the Mid-Atlantic, Northeast and Western seaboards of the CONUS. A weather reporting station located in close proximity to each route segment was identified in order to obtain a dataset of historical winds and atmospheric conditions.

A section of the aviation chart depicting the route segments utilized for the analysis and their associated weather station locations is presented in **Figure 5**.

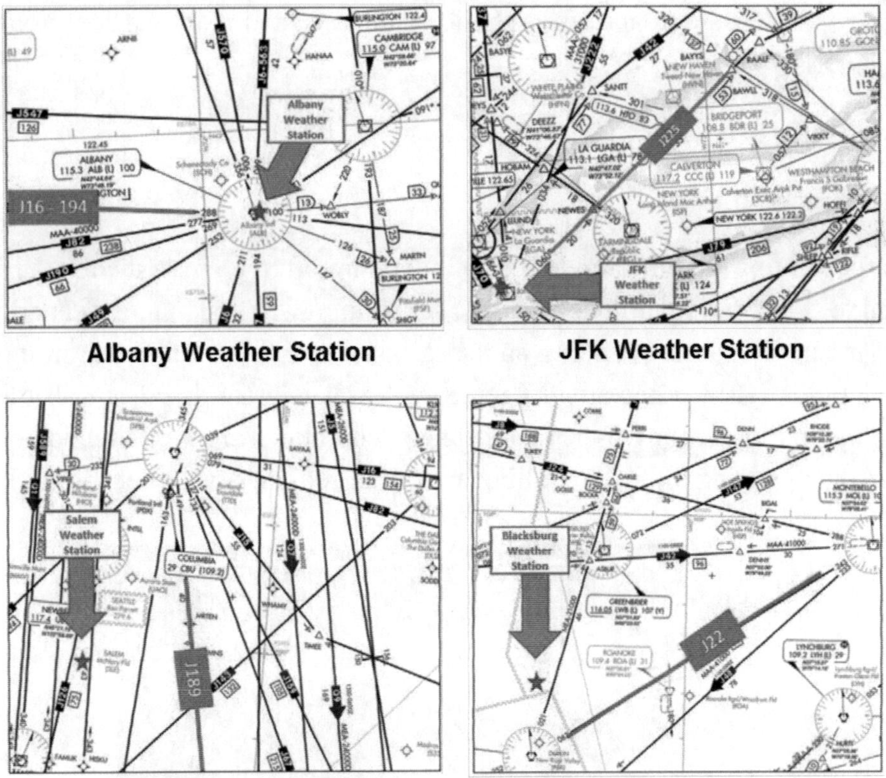

Albany Weather Station JFK Weather Station

Salem Weather Station Blacksburg Weather Station

Figure 5: Route and Weather Station Selection

Radiosonde weather balloon data was obtained for pressure altitudes of 200, 300, 400 and 500mb for a reporting period of 10 years. In the example below, a crosswind component of 5KTS was used as the minimum velocity to transport wake away from the operating route. The data sets were parsed to determine the number of instances when reported winds resulted in crosswind components less than the calculated value. **Table 1** presents a summary of the recorded wind data reported as the percentage of instances when the crosswind component of the reported winds was calculated as 5 knots or less.

Pressure Reading (mb)	Altitude (ft)	Station			
		ALB	Blacksburg	JFK	Salem
500	18,300	17.0%	16.8%	9.6%	10.7%
400	23,500	14.6%	13.9%	9.7%	8.8%
300	30,000	11.5%	11.9%	7.5%	7.7%
200	38,700	12.3%	9.6%	6.1%	6.8%
Average Across All Altitudes		**13.8%**	**13.0%**	**8.2%**	**8.5%**

Table 1 Radiosonde Readings - Crosswind ≤ 5KTS

Besides crosswinds, same route wake encounters are also dependent on aircraft positional uncertainty, operating speed and atmospheric conditions, and require a comprehensive evaluation to accurately assess likelihood.

Wake Encounter Scenarios

Aircraft Operating In Trail

For aircraft operating en route and in trail of other aircraft, the greatest wake turbulence risks are anticipated to occur during altitude changes. A common wake encounter scenario is presented on Figure 6. An aircraft pair is operating on the same route and the trailing aircraft is executing a change in altitude. The leading aircraft generates a wake vortex that descends through the altitude of the trailing aircraft. When atmospheric conditions are such that do not drift the wake away from the operating route, the trailing aircraft is at higher risk of encountering the vortex field.

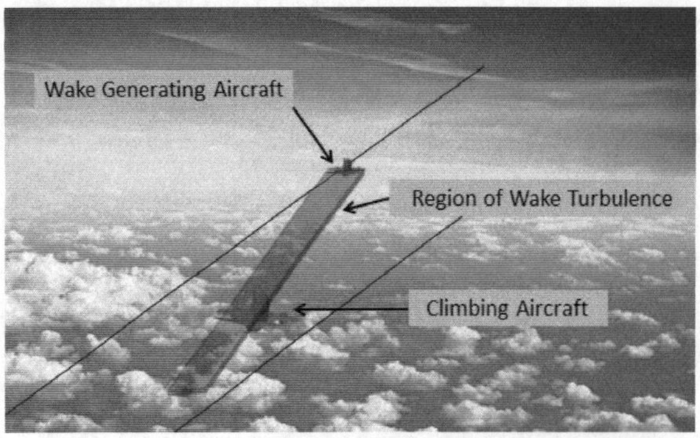

Figure 6: In-Trail Operating Scenario

Head-On Pass Scenario

A head-on pass scenario is one in which two aircraft are operating in opposite directions on the same route with a net altitude difference of 1,000 feet. Two of the many factors that contribute to the likelihood of a wake encounter during a head-on passing scenario are the aircraft pair vertical separation and the descent rate of the wake turbulence. While the cross-track wind velocity remains a mitigating factor, the time it takes for the vortex to descend to the altitude of the lower aircraft will determine the precise cross-track wind velocity needed transport the wake turbulence off route. Research has shown that, in general, wake descent rates vary between aircraft types and are also affected by atmospheric conditions.

En Route Wake Mitigation

In scenarios where cross-track wind threshold does not provide adequate wind mitigation of the wake turbulence, other options to reduce the risk of an encounter may include increasing along-track separation, altitude changes or lateral offsets. In-trail same altitude separation increases would require a speed change for one or both aircraft and may not be operationally feasible or desirable. Altitude changes or increased vertical separation can avoid areas of wake turbulence or provide additional time for wake transport or decay to occur. In some cases, altitude changes may not be possible due to traffic conflicts. A third option, lateral offsets, may be an option for applying mitigation to reduce en route wake turbulence risk.

A lateral offset is a maneuver that establishes an aircraft on a parallel route or trajectory at some fixed distance from the primary route. Offset distances may be in increments of as little as 0.1NM to in some cases 5.0NM. The offset route is defined by the flight crew using the aircraft's Flight Management Computer (FMC) and is a parallel clone of the original route. The ability of FMC's to perform an incremental distance offset is present in most "modern" FMC's, although some early first-generation FMC's be limited to offset distances in whole increments of 1.0 NM.

The assessment of the current wind conditions is an important aspect of the use of a lateral offset. Guidance in the ICAO Doc 4444 PANS for implementation of the Strategic Lateral Offset Procedure (SLOP) states that the "decision to apply a strategic lateral offset shall be the responsibility of the flight crew". The use of "standardized" maneuvers as part of the offset procedure also serves to enhance the predictability of the maneuver and better ensure that the desired separation from wake turbulence is maintained. These offsets can be classified as two basic types; those that are procedural and conducted without regard to wind conditions, and those that are wind dependent and consider cross-track winds.

Procedural Offsets

The most common procedural offset is the Standard Lateral Offset Procedure (SLOP), which is currently used in oceanic airspace. The SLOP authorizes an offset in incremental distances of

0.1NM to a maximum limit that is dependent on the route centerline spacing. For areas where the route centerline spacing is 23NM or more, the maximum limit of the offset distance is 2NM. In areas where the centerline spacing is at least 6NM but less than 23NM, the maximum offset distance is limited to 0.5NM. The SLOP is always made to the right of course direction and without regard for wind direction or velocity. Adverse crosswinds transport wake in the direction of an offset. Under these wind conditions, the use of the SLOP by crews expecting a wake mitigation benefit, may actually be placing the aircraft in a region with the highest probability of encountering wake turbulence.

Wind Dependent Offsets

A wind dependent offset is one that also utilizes either an incremental or whole integer offset distance, but is performed in the upwind direction and enables the cross-track wind to provide additional separation between the aircraft and wake turbulence. A standardized offset is a procedure that combines a standard set of maneuvers and a wind based "intelligent" offset direction over a prescribed offset distance. A standardized offset for an in-trail climb is depicted in **Figure 7**.

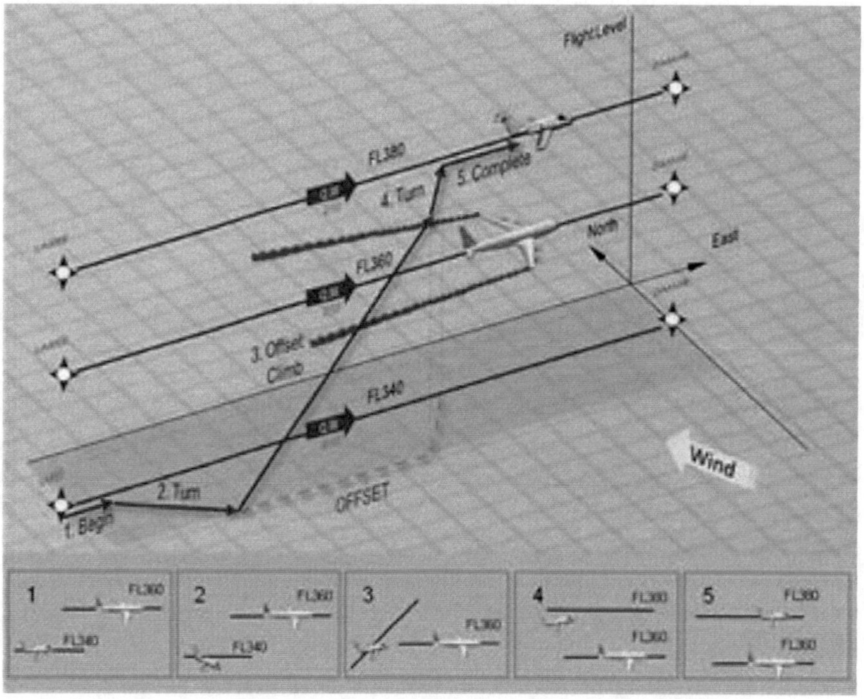

Figure 7: Wind Dependent Standardized Offset

Summary and Conclusion

Consistent reports from pilot and controllers of wake turbulence encounters at high altitudes suggest that en route wake turbulence is a recurrent safety concern and will need to be considered when reducing aircraft separation and increasing NAS capacity. While a significant amount of research and wake turbulence data has been collected at low altitudes near airports, the lack of measured data at high altitudes has limited the understanding of aircraft vortex behavior during the en-route phase of flight. The FAA in collaboration with NRC has collected high altitude wake turbulence data to broaden the understanding of the risks associated with these vortex systems.

As some of the en route wake reports suggest, the analysis of high altitude wake vortex data has demonstrated that, in some instances, significantly strong vortex cores exist at distances in trail well beyond current separation minimums. In some cases, vortex fields were measured beyond 20 nautical miles in-trail of the generating aircraft. The strength of these vortices, as well as transport and decay rates is highly dependent on atmospheric conditions, limiting the potential to predict the risk associated with a wake vortex system. This research calculated those atmospheric conditions for which there is a higher the potential for a wake encounter, and evaluated historical wind data to estimate the frequency of occurrence of such events.

Finally, implementing wake mitigation strategies may allow increasing NAS capacity without compromising safety. These strategies include along track separation increases, altitude changes or lateral offsets. While increasing along track separation or altitude changes may reduce the likelihood of an encounter, it doesn't contribute to increasing capacity. Lateral offsets, such as the currently used SLOP take best advantage of the airspace. However this specific offset does not account for wind conditions and is authorized to only the right side of the direction of the route, which can increase the potential for a wake encounters when the crosswind transports the wake in the direction of the offset. Wind dependent offsets are concluded as a potential mitigation to reduce the likelihood of wake encounters and increase capacity, as it can utilize small incremental offsets that remain well within route containment boundaries.

AUTHOR INDEX